PLAGUES

Throughout history plagues have inflicted misery and suffering on humankind. Traces of these catastrophes are echoed in our literary past, and passed on through generations in our genes. We often associate plagues with epidemic diseases, yet plagues are much more. Featuring essays arising from the 2014 Darwin College Lectures, this book examines the spectrum of tragic consequences of different types of plagues, from infectious diseases to over-population and computer viruses. They analyse the impact that plagues have had on humanity and animals, and their threat to the very survival of the world as we know it. Each essay takes a typically diverse approach to its theme, ranging through history, medicine, ecology, archaeology, molecular biology, economics, computer viruses and biblical metaphor. This engaging and timely collection challenges our understanding of plagues, and questions if plagues are the manifestation of nature's checks and balances in light of human population growth and our impact on climate.

JONATHAN L. HEENEY is a Fellow of Darwin College and Professor of Comparative Pathology in the Department of Veterinary Medicine, and Head of the Laboratory of Viral Zoonotics at the University of Cambridge.

SVEN FRIEDEMANN is a former Schlumberger Fellow of Darwin College and Feodor Lynen Fellow of the Alexander von Humboldt foundation at the Cavendish Lab, Department of Physics in Cambridge, and is now a lecturer in the School of Physics at the University of Bristol.

THE DARWIN COLLEGE LECTURES

These essays are developed from the 2014 Darwin College Lecture Series. Now in their twenty-ninth year, these popular Cambridge talks take a single theme each year. Internationally distinguished scholars address the theme from different points of view from a wide range of academic disciplines.

Subjects Covered in the Series Include

28 PLAGUES
ed. Jonathan L. Heeney and Sven Friedemann
pb 9781316644768

27 FORESIGHT
ed. Laurence W. Sherman and David Allan Feller
pb 9781107512368

26 LIFE
ed. William Brown and Andrew Fabian
pb 9781107612556

25 BEAUTY
ed. Lauren Arrington, Zoe Leinhardt and Philip Dawid
pb 9781107693432

24 RISK
ed. Layla Skinns, Michael Scott and Tony Cox
pb 9780521171977

23 DARWIN
ed. William Brown and Andrew Fabian
pb 9780521131957

22 SERENDIPITY
ed. Mark de Rond and Iain Morley
pb 9780521181815

21 IDENTITY
ed. Giselle Walker and Elizabeth Leedham-Green
pb 9780521897266

20 SURVIVAL
ed. Emily Shuckburgh
pb 9780521718206

19 CONFLICT
ed. Martin Jones and Andrew Fabian
hb 9780521839600

18 EVIDENCE
ed. Andrew Bell, John Swenson-Wright and Karin Tybjerg
pb 9780521710190

17 DNA: CHANGING SCIENCE AND SOCIETY
ed. Torsten Krude
hb 9780521823784

16 POWER
ed. Alan Blackwell and David Mackay
hb 9780521823717

15 SPACE
ed. Francois Penz, Gregory Radick and Robert Howell
hb 9780521823760

14 TIME
ed. Katinka Ridderbos
hb 9780521782937

13 THE BODY
ed. Sean Sweeney and Ian Hodder
hb 9780521782920

12 STRUCTURE
ed. Wendy Pullan and Harshad Bhadeshia
hb 9780521782586

11 SOUND
ed. Patricia Kruth and Henry Stobart
pb 9780521033831

10 MEMORY
ed. Patricia Fara and Karalyn Patterson
pb 9780521032186

9 EVOLUTION
ed. Andrew Fabian
pb 9780521032179

8 THE CHANGING WORLD
ed. Patricia Fara, Peter Gathercole and Ronald Laskey
pb 9780521283281

7 COLOUR: ART AND SCIENCE
ed. Trevor Lamb and Janine Bourriau
pb 9780521499637

6 WHAT IS INTELLIGENCE?
ed. Jean Khalfa
pb 9780521566858

5 PREDICTING THE FUTURE
ed. Leo Howe and Alan Wain
pb 9780521619745

4 UNDERSTANDING CATASTROPHE
ed. Janine Bourriau
pb 9780521032193

3 WAYS OF COMMUNICATING
ed. D. H. Mellor
pb 9780521019040

2 THE FRAGILE ENVIRONMENT
ed. Laurie Friday and Ronald Laskey
pb 9780521422666

1 ORIGINS
ed. Andrew Fabian
pb 9780521018197

Plagues

Edited by *Jonathan L. Heeney* and *Sven Friedemann*

Darwin College, Cambridge

CAMBRIDGE
UNIVERSITY PRESS

University Printing House, Cambridge CB2 8BS, United Kingdom

One Liberty Plaza, 20th Floor, New York, NY 10006, USA

477 Williamstown Road, Port Melbourne, VIC 3207, Australia

4843/24, 2nd Floor, Ansari Road, Daryaganj, Delhi - 110002, India

79 Anson Road, #06-04/06, Singapore 079906

Cambridge University Press is part of the University of Cambridge.

It furthers the University's mission by disseminating knowledge in the pursuit of education, learning, and research at the highest international levels of excellence.

www.cambridge.org
Information on this title: www.cambridge.org/9781316644768

First published 2017

Printed in the United Kingdom by Clays, St Ives plc

A catalogue record for this publication is available from the British Library.

Library of Congress Cataloging-in-Publication Data
NAMES: Heeney, Jonathan L. (Jonathan Luke), 1957– editor. | Friedemann, Sven, 1980– editor.
TITLE: Plagues / edited by Jonathan L. Heeney and Sven Friedemann.
OTHER TITLES: Plagues (Heeney) | Darwin College lectures ; 27.
Description: Cambridge, United Kingdom ; New York, NY : Cambridge University Press, 2017. | Series: Darwin College lectures ; 27 | Includes bibliographical references and index.
IDENTIFIERS: LCCN 2016049433 | ISBN 9781316644768 (paper back : alk. paper)
SUBJECTS: | MESH: Communicable Diseases | Epidemics | Hemorrhagic Fever, Ebola
CLASSIFICATION: LCC RA649 | NLM WC 100 | DDC 614.4/9–dc23 LC record available at https://lccn.loc.gov/2016049433

ISBN 978-1-316-64476-8 Paperback

Dedicated to:

Professor Sir Patrick Sissons (1945-2016).
Virologist, Physician, Clinical Scientist,
Regius Professor of Physic, Fellow of Darwin College.
An advocate of clinical infectious disease research,
the Cambridge in Africa project, and Global Health.

Professor Marian C Horzinek (1936-2016).
Virologist, Veterinary Scientist,
Eminent scholar and mentor of a generation of European
Virologists. An advocate of One Health, One Medicine.

Contents

Figures

Tables

Notes on Contributors

Leszek Borysiewicz FRS studied medicine at the Welsh National School of Medicine and received a PhD from the Royal Postgraduate Medical School. He is the 345th Vice-Chancellor of the University of Cambridge and was previously the Chief Executive of the UK's Medical Research Council (2007–10). From 2001 to 2007 he was Principal of the Faculty of Medicine and Deputy Rector at Imperial College London, responsible for the overall academic and scientific direction of the institution. Whilst at the University of Wales in Cardiff he led a research team that carried out pioneering work on human papillomavirus vaccines, in particular a therapeutic vaccine for cervical cancer. He was knighted in 2001 for services to medical research and education. He was a founding Fellow of the Academy of Medical Sciences and is a Fellow of the Royal Society.

Christopher Dobson FRS studied chemistry for his BA and Doctorate at Oxford. He is the John Humphrey Plummer Professor of Chemical and Structural Biology at the University of Cambridge and Master of St. John's College. He is a Fellow of the Royal Society and of the Academy of Medical Sciences and a Foreign Associate of the US National Academy of Sciences. He has received numerous awards, including both the Davy and Royal Medals of the Royal Society, and achieved international recognition for his research on the nature of protein misfolding and its links to disease. His work has provided a fundamentally new view of the origins and means of progression of many of the most debilitating and increasingly prevalent 'plagues' of the modern era, ranging from Alzheimer's disease to Type II diabetes.

Mary Dobson studied geography at Oxford for her BA and doctorate degrees and is now a historian of medicine with wide-ranging interests in the plagues of the past. She has held a number of research fellowships, including a Harkness Fellowship at Harvard University, and was formerly Director of the Wellcome Unit for the History of Medicine at the University of Oxford, and a Fellow of Green Templeton College. She is author of a variety of publications, including a groundbreaking monograph, *Contours of Death and Disease in Early Modern England* (Cambridge University Press, 1997). Her books for general audiences include *Disease: the Extraordinary Stories Behind History's Deadliest Killers* and *The Story of Medicine: From Bloodletting to Biotechnology*. Her latest book, *Murderous Contagion*, has recently been released.

Stephen Emmott studied experimental psychology at the University of York and was awarded a PhD in Computational Neuroscience at the University of Stirling. In 2009 he was appointed by the UK Science Minister as a trustee of The National Endowment for Science, Technology & the Arts. He is a visiting professor of biological computation at University College London, visiting professor of computational science at the University of Oxford, and a Distinguished Fellow of The National Endowment for Science, Technology & the Arts. He has headed Computational Science, Microsoft, where he leads an interdisciplinary research programme centred on Microsoft's Computational Science Laboratory in Cambridge.

Jonathan L. Heeney studied Veterinary Medicine in Ontario, Canada, thereafter receiving a doctorate in Pathology. Subsequently he undertook research for his PhD in viral immunopathology at the NIH (Maryland) then took up a Fellowship in molecular and comparative pathology at the Stanford School of Medicine (California). In the 1990s Jonathan established the Laboratory of Viral Pathogenesis in the Netherlands where he studied viral infections of immunocompromised hosts and pioneered a number of candidate vaccines for the prevention of HIV/AIDS and Hepatitis C amongst others. He established an international series of meetings and think tanks for vaccine design based on immune correlates. In 2007 he was elected Professor of Comparative Pathology at the University of Cambridge where he established the Laboratory of Viral Zoonotics, a lab dedicated to the study of viral diseases transmitted from animals to humans. He has published widely on globally important human diseases, from AIDS to Ebola and their zoonotic origins in animals.

Mikko Hypponen is the Chief Research Officer of F-Secure in Finland and a columnist. He has fought the biggest virus outbreaks in the net, including Loveletter, Conficker and Stuxnet. His TED Talk on computer security has been seen by almost a million people and has been translated into more than thirty-five languages. He has addressed the EU Parliament and his columns have been published in the *New York Times*, *Wired*, CNN and the BBC. Mr Hypponen was selected among the fifty most important people on the web by *PC World* magazine. *Foreign Policy* magazine included him in the list of 'Top 100 Global Thinkers'. Mr Hypponen sits on the advisory boards of the ISF and the Lifeboat foundation.

Angela McLean FRS studied mathematics at Oxford and earned a PhD in biomathematics from Imperial College, London. After a brief spell in the City she joined the Mathematical Biology Group at the National Institute for Medical Research at Mill Hill then returned to Oxford as a Royal Society University Research Fellow. In 1994 she went on secondment to the Institut Pasteur in Paris to work on the population dynamics of murine lymphocytes in the immunology department there. In 1998 she became Head of Mathematical Biology at the BBSRC's Institute for Animal Health. Since 2008 she has been a Senior Research Fellow at All Souls College in Oxford. She was elected to the Royal Society in 2009, and was awarded the Royal Society's Gabor Medal in 2011.

Ian Morris studied ancient history and archaeology at the University of Birmingham and earned his PhD in classics at Cambridge. He is the Jean and Rebecca Willard Professor of Classics and Archaeology at Stanford University. He taught at the University of Chicago before moving to Stanford. His most recent books are *Why the West Rules—For Now: The Patterns of History and What They Reveal About the Future*, *The Measure of Civilization: How Social Development Decides the Fate of Nations* and *War! What is it Good For? Violence and the Progress of Civilization, from Primates to Robots*. He has directed archaeological excavations in Italy and Greece and is a Corresponding Fellow of the British Academy.

Steven J. O'Brien studied biology at St Francis College, Pennsylvania, and earned his PhD in genetics at Cornell University. He is a molecular geneticist who uses the tools of molecular biology to help protect endangered species and understand devastating diseases such as cancer and AIDS. He joined the National Cancer Institute, National Institutes of Health as a post-doc in 1971 and there served as Founder and Chief of the Laboratory of Genomic Diversity from 1986–2011. In December 2011, he joined the Theodosius Dobzhansky Center for Genome Bioinformatics, St. Petersburg State University, as Chief Scientific Officer. His research interests and expertise span human and comparative genomics, genetic epidemiology, HIV/AIDS, retrovirology, bioinformatics biodiversity and species conservation.

He revealed the importance of CCR5 delta 32, one of more than twenty human AIDS restriction genes, and one which imparts natural resistance to HIV infection and/or the development of AIDS.

Rowan Williams studied theology at Cambridge and earned his doctorate at Oxford. He became Tutor and Director of Studies at Westcott House, Cambridge, and is currently Master of Magdalene College Cambridge. After ordination in Ely Cathedral, and serving as Honorary Assistant Priest at St George's Chesterton, he was appointed to a university lectureship in Divinity. In 1984 he was elected a Fellow and Dean of Clare College. During his time at Clare he was arrested and fined for singing psalms as part of the CND protest at Lakenheath airbase. He was enthroned Archbishop of Wales in 2000 and Archbishop of Canterbury in 2003. He holds several honorary doctorates and is a Fellow of the British Academy. Dr Williams is a noted poet and translator of poetry, and, apart from Welsh, speaks or reads nine other languages. He learnt Russian in order to read the works of Dostoevsky in the original. He has published studies of Arius, Teresa of Avila and Sergii Bulgakov, together with writings on a wide range of theological, historical and political themes.

Foreword

We associate plagues with epidemic diseases. The Darwin lecture series on *Plagues* was, by tragic serendipity, unwittingly launched on the eve of an erupting epidemic of humankind's most feared viral disease, Ebola. Little did we know that 2014 would be the start of the most serious single Ebola outbreak of our recorded history. It highlighted how plagues of such severe disease affect individuals in all layers of society, illustrating the frailty of human life with the unraveling of fragile health care systems, basic infrastructure, transportation and trade, leaving national economies on the brink of collapse.

Yet, the word 'plagues' has many more meanings. We explored these through the multidisciplinary approach used by the Darwin College Lecture Series, enriched by views of prominent scholars from very different fields. In this book the theme 'plagues' is portrayed not only by international experts of infectious diseases (vaccinology, biochemistry and medicine), but also from scholars in digital security, computational science, mathematical biology, ecology, genetics, history, economics, literature and theology. The result is a comprehensive spectrum of essays on plagues, from their impact on humanity to the animal kingdom, all bundled together in this one very unique book. It is another successful culmination of diverse perspectives on one such theme that the Darwin Lectures so uniquely weaves into one intellectually challenging topic.

Throughout history plagues have inflicted misery and suffering on humankind. Traces of these catastrophes are echoed in our literary past and, intriguingly, also encrypted and passed on through generations in our genes. Notably for the future of human population growth, the expansion of our species is challenging and changing the very environment we and other species require for survival on earth. We are also reminded that it may not just be the infectious plagues but also non-infectious ones that will impact this planet's species as this generation ages. The ever-expanding human population and our impact on climate threatens the delicate balance of ecosystems, making us a human plague to this fragile planet. Are plagues the manifestation of nature's checks and balances? This collection is a thought-provoking product of the topics of the 2014 Darwin College Lecture Series delivered by internationally renowned scholars, known not only for their outstanding work and literary talents, but for their ability to communicate the diverse facets and impacts of plagues.

Preface

Human history, evolution and even our religious and cultural practices have been greatly influenced by plagues of infectious diseases. The Zika virus epidemic in South America has left a generation of children with microcephaly and raised concerns about infections acquired during large international gatherings such as the 2016 Olympic Games in Rio de Janeiro. The year 2014 will be remembered as the year that a highly contagious and deadly viral disease, Ebola, for the first time moved into large urban populations and quickly became a plague of high mortality and affliction. It directly affected three West African nations and their neighbouring countries, while threatening other nations around the globe. The 2014 Darwin College lecture series 'Plagues' had been conceived 18 months earlier, based not on a premonition of an Ebola epidemic, but on the inevitability of an outbreak of one of the many emerging and re-emerging infectious diseases from around the world. The 'Plague' lecture series and this book transcend the medical aspects of plagues by remaining true to the tradition of the multidisciplinary engagement of a single theme examining plagues of different types as well as those affecting other species in addition to humans. *Plagues* draws not only from the biological and medical sciences but also a spectrum of diverse fields such as those which infect our digital global network, threaten our financial institutions, and which have even influenced our language and literature. Intentionally, the sequence and the content of the chapters diverge from the platform lectures, which were limited in time and scope, thus now providing the reader with greater depth, detail and a broader conceptual perspective.

Due to an unfortunate coincidence, the 'Plague' lecture series was immediately followed by confirmation of the largest outbreak of one of humankind's most feared infections, Ebola. We have therefore most appropriately included a chapter 'Ebola, the Plague of 2014/15'. This large-scale epidemic engulfed West Africa and threatened International borders. It took more lives than all other recorded Ebola outbreaks before it and dominated international headlines for two years. This contemporary plague began unrecognised with the death of a two-year-old boy in December 2013, a month before the Darwin College lecture series on plagues would start. Tragically, because Ebola was unknown to West Africa it was misdiagnosed until late March 2014. With increasing momentum the number of cases of haemorrhagic fever began to accumulate as the infection was transported from rural settings in Guinea to highly populated urban centres in the neighbouring countries Sierra Leone and Liberia. Only in March 2014, a week after the conclusion of the Darwin Lectures on 'Plagues', was the diagnosis of Ebolavirus confirmed, three months after the first case occurred. In retrospect, many of the broad social/anthropological, public health and economic issues that arose during this West African Plague of 2014 were themes that the 'Plague' series had encompassed and are provided in greater detail in this book.

Distinct from the lectures, these chapters follow a progression from the historical to the medical and evolutionary perspectives through to the inseparable dynamics of animal and human infections, considering as well the social, economic and computational examples of how plagues of different kinds have afflicted all aspects of our society, including the impact of the ever-expanding human population on our planet. Finally, this book concludes with an example of the impact that plagues have had on our literary heritage.

Plagues of different types have been recorded in some of the earliest historical records depicting or describing devastating afflictions, such as the Biblical plagues, or the Plague of Athens and the Justinian Plague. To co-author 'Plagues and History' no one other than the medical historian Mary Dobson could have been better placed to take us through the history of plagues and how they have impacted humankind. She examines chronologically plagues that have influenced our history: pharaohs, kings and queens and their armies alike that have been felled by contagious diseases. She describes how plagues ranging from 'the Black Death' (*Yersinia pestis* 'The Plague'), to smallpox, typhus and cholera, have shaped our history. Infectious diseases followed European explorers and settlers to the New World and killed or afflicted huge numbers of the indigenous peoples of the Americas and Australia. Now much more is known about the biology of the infectious agents and the events which led to the emergence of plagues of the modern world; such as influenza (the 1918 Spanish Flu pandemic), SARS, TB and HIV. Here in a fantastic partnership Christopher Dobson illustrates the compelling case for a growing silent epidemic of plagues of the future. He describes the stealth-like epidemic of non-infectious plagues of neurodegenerative diseases. These diseases are growing in prevalence as populations of developed countries survive longer. Interestingly, statistics suggest that cases of dementia and Alzheimer's will soon overwhelm health care resources and represent major challenges to medicine and elderly care.

As plagues have emerged and re-emerged they have evoked a human response. In 'Plagues and Medicine' Leszek Borysiewicz illustrates how the medical profession eventually evolved effective measures to deal with infectious diseases; but points out that such measures are frequently not adapted without a degree of public scepticism. The public health practice of 'quarantine' as a response of the return of the Black Death first proved to be effective in the seventeenth century and slowly became widely applied in the eighteenth century. So effective was the practice of quarantine for human diseases that when the 'Cattle Plague' (Rinderpest) descended upon Italy in the early 1700s, the Pope decreed the quarantine principles of his medical advisor be followed with strictest adherence to prevent the transmission of the Rinderpest virus (viruses then were an unknown entity). The application of quarantine practices was proven to be effective for human and animal plagues, a practice which interestingly led to the establishment of the first school of veterinary medicine.

As illustrated in the 2014 Ebola outbreak, public health measures, quarantine and travel restrictions proved difficult to enforce as cultural and religious belief and public mistrust often departed from best medical practice. Sir Leszek makes this point eloquently using the historical example of vaccination and the resistance that evolved from the anti-vaccination movement consisting not only of the lay public, theologians and politicians, but even some of the most revered scientists of their day (such as Alfred Russell Wallace). And as he points out, even after the remarkable achievement of the eradication of smallpox by vaccination in 1980, there remains today a vocal minority who are sceptical and reluctant, risking the health of their children and those who they contact. Using the measles, mumps, rubella vaccine example, where unfounded rumours of autism had decreased voluntary vaccination, he makes the strong argument that improved vaccine uptake in the future will depend on understanding and addressing the complex issues impacting on public health perceptions in society, as well as finding the balance between individual rights and societal needs.

Where do modern plagues come from and how do they arise? As Angela McLean explains in 'The Nature of Plagues' they can be new agents or known infections spreading into new populations. She discusses new infectious diseases which have changed species and or mutated to such a degree that they escape immune recognition in previously exposed populations. The 'Nature of Plagues' becomes obvious when examining how infectious diseases spread to new geographically separate regions where there is no immunity in human or animal populations. Many complex factors influence the emergence of new or re-emerging infectious diseases. Using contemporary examples from June 2013 to May 2014, Angela McLean illustrates the different events that lead to the emergence of infectious disease. Middle East respiratory syndrome coronavirus (MERS-CoV) is a newly recognised human coronavirus transmitted from camels. Many camels carry antibodies to this virus and infected juveniles are suspected to be the reservoir of infections for this zoonotic infection (transmitted from animals to humans). Human to human transmissions have been reported but secondary cases are infrequent, unlike its highly contagious cousin the SARS-CoV. One major concern is that the MERS-CoV may evolve, giving rise to variants that become highly transmissible from humans to humans. Close monitoring of the MERS-CoV epidemic is warranted, as the recent outbreaks have illustrated. McLean's 2013/2014 chronology of outbreaks of infectious disease illustrates the frequency and diversity of different infectious agents and the often unexpected patterns of outbreaks. One of the most notable outbreaks of recent years was the steady spread of Zika, Chikungunya and Dengue viruses to the New World where they have rapidly affected many countries in South America, and have since been steadily marching north into the United States.

The cases of microcephaly associated with Zika infection during pregnancy have raised particularly complex societal debates regarding religion, gender, abortion and disability rights. Two factors relate to the rapid emergence of these viral diseases in this previously unaffected part of the world: 1) the presence of two aggressive species of *Aedes* mosquitos, which carry them and 2) an immunologically naïve and thus fertile human population providing these mosquitos with a growing human blood supply of virus-infected meals.

Relatively unnoticed was the resurgence of polio, which was overshadowed by the media's high profile coverage of the Ebola epidemic in West Africa (Chapter 1). Unfortunately, despite Polio being identified as an emergency of international concern, global vaccine coverage has not yet been realised even though there is an effective vaccine (Chapter 3). The 2013/2014 examples of new emerging and re-emerging disease dynamics portray the differences in the natural history of infections caused by shifts in the demographics of epidemics, changes in geographic distribution of animals, biting vectors and human perceptions; all factors which may favour increases in infectious disease transmission and spill-over events to other species.

Are there more infectious diseases than in the past and are we equipped to identify them? Possibly, but as McLean correctly argues there are probably more animal to human transmission events than in the past simply because there are more humans and they are certainly more widely distributed around the globe, penetrating foreign ecosystems like never before.

If human and animal populations have been constantly challenged and depleted by deaths due to infectious diseases, what genes have the survivors of these pandemics passed on to future generations? Are some of us now better equipped genetically to survive certain infectious disease threats? In 'Plagues, Populations, and Survival' S.J. O'Brien reflects on his experiences from the early beginnings of the genomics era. When geneticists study today's animal species in the light of Darwinian evolution, they consider disease resistance acquired by ancestral survivors that are passed on to subsequent generations through the process of natural selection. O'Brien begins with the example of the SARS epidemic and a novel human coronavirus that was acquired from civet cats sold for human consumption in Guangdong province in China (ultimately the origin of the SARS virus proved to be horseshoe bats). As O'Brien points out, coronaviruses were best known to veterinarians for the serious diseases they occasionally caused in pigs and cats. It was the devastating loss of a large number of captive cheetahs in Oregon that highlighted how vulnerable a genetically uniform and immunologically naïve population could be to new infectious agents. In that case a common domestic cat coronavirus had spread through a non-indigenous population of a highly susceptible and genetically uniform population of cheetahs, causing high morbidity and 60 per cent mortality.

In another example of cross-species transmission, O'Brien describes the human immunodeficiency viruses. HIV/AIDS had its origins in non-human primates in Africa, and ultimately, in the case of HIV-2, was acquired from infected Sooty Mangabey monkeys, and in the case of HIV-1 from the consumption of SIV-infected chimpanzees as bushmeat. But AIDS and AIDS-causing lentiviruses are not only found in primates (including humans). O'Brien points out that domestic and wild cats carry an HIV-related lentivirus called Feline immunodeficiency virus (FIV). He and others had showed that FIV had been transmitted historically to other large cat families. AIDS caused by FIV in domestic cats is not always as pathogenic as untreated HIV-1 in humans, and without co-factors such as feline leukaemia virus or other opportunistic infections AIDS in cats is not usually manifested. Different strains of FIV also seem to be more pathogenic than others. O'Brien illustrates this in the case of a dog morbillivirus infection that caused severe disease when transmitted to lions carrying a specific strain of FIV.

Notably, it was the human genome project and the sequencing of genomes from AIDS-susceptible or AIDS-resistant humans where the identification of resistance genes provided the greatest wealth of information. O'Brien explains that certain genetic mutations found in some of us can protect against HIV. An intriguing mystery which he reveals is that this genetic mutation seems to have been selected in Caucasians long before the relatively recent HIV lentivirus spread globally. While the 'Plague' that selected for this particular mutation is still a mystery, there is accumulating evidence that infectious diseases have helped shape our DNA landscape by causing the death, and hence the loss, of many gene variants which caused susceptibility or had no beneficial effect during an epidemic. Following catastrophic losses of individuals from populations, we have inherited the genes of the survivors, some of which carry genetic changes that may help our very survival if again faced with the same infectious causes of these historic plagues.

What are the consequences of plagues on societies and national economies once they have come and gone? In 'Plagues and Socioeconomic Collapse', Ian Morris undertakes a fascinating analysis of historical plagues, asking how they impacted on the fortunes of mankind. Do plagues cause socioeconomic collapse? He notes that the conclusions are not always what one would expect, and the acute and long-term outcomes are often different. He chooses as his index case the Black Death of the fourteenth century, the archetypical plague in magnitude and duration, having killed more than a third of the population of Europe, China and the Middle East. In Europe its consequences were far reaching over the following century. Yes, the Black Death did cause socioeconomic collapse, but not as severe or prolonged as the Antonine Plague, which caused a multi-century collapse. Surprisingly, a century later, the long-term consequences of the Black Death were generally positive, with the shift from feudalism to capitalism in the West triggered by the change in demographics. The recovery from the plague provides an interesting comparison of the impact on social and political structures and the outcome of recovery, while major historical events such as changes in trade routes played background roles. This is a fascinating chapter that in the context of the evolving historical events provides a number of unexpected conclusions and sobering reminders that increased human mobility fuels global epidemics. How will our global societies and evermore interlinked economies cope with future plagues?

Our lives and indeed our world are now dependent on the digital networks that control our communication, the movement of food, the transportation of people, global financial markets and even our personal bank accounts. We live in a highly interconnected fast-moving digital world. Our personal details, bank accounts, social networks and daily lives are influenced by events interconnected through the Internet. Businesses, doctors' offices, hospitals, the stock market, news rooms and even intelligence agencies and war rooms are now dependent on vast computer networks. How susceptible are we to a digital or electronic plague? In his chapter 'Silicon Plagues' Mikko Hypponen describes the emergence of computer viruses, those that started as intellectual challenges from rascally young minds, to those with deliberate criminal intent aimed at defrauding and stealing personal assets to robbing banks or corporate secrets while constantly challenge national security agencies operating large-scale surveillance. There is a cold war against cybercrime that is ongoing daily. Professional hackers use ingenious ways to find their way into computer databases, to steal or distribute sensitive data. Personal, corporate, and now even large-scale cyberattacks have occured during recent election campaigns and continue on a daily basis. What and when will we be faced with a silicon virus that will secretly infect and plague our computers, threaten our livelihood, the stock market, our pensions, or the security of nations? The time for this new affliction is upon us, but for how long can we defend against potentially devastating consequences.

The natural world of planet earth has been changed by the success of the human race. Climate change may be the tip of the iceberg as we change the ecosystems and threaten the survival of species with whom we share this planet. In 'The Human Plague', Stephen Emmott eloquently illustrates the hard realities of the affliction the ever expanding human population has brought upon the natural world and the fragile global ecosystem as we know it. How many billions of humans can the world sustain and at what consequence? Have we already triggered global changes that may be catalytic and accelerated the consequences of global warming beyond the point of no return? What is the impact of megacities over 20 million people, the expansion of agricultural land and the unbridled deforestation of precious rainforests? What are the consequences for the earth's biosphere that we require for our own existence? How far can it be pushed before mankind eradicates itself? Is the Human Plague the final plague?

As Rowan Williams points out in the final chapter on 'Plague as Metaphor' the word 'plague' has come to imply something greater and more sinister than an accident, more akin to a disaster or catastrophe. Historically, we attributed divine interventions to plagues. It is with great skill that this masterful discussion of the metaphoric use of plague is crafted by comparison of different literary works.

Following the gloom portrayed by the revelations of Emmott's human population projections and the climatic turmoil the human race appears to have put in motion, a plague of divine agency may someday be upon us. Will we recognise the accumulating evidence and the consequences in sufficient time to collectively avert the next plague, or will we as the human race ultimately be victims of our own success?

Jonathan L. Heeney & Sven Friedemann

Acknowledgements

This book would not have come together without the lecture series that inspired it and the many members of Darwin College who facilitated the 8-week event. We are in appreciation to Richard and Anne King for their generous financial support. Our special thanks to all who assisted with proof reading, with special note to Belinda Cheney, whose comments, insights and advice were invaluable. Janice Jermy's and Janet Gibson's patience, dedication, and friendly and meticulous organisation have been central to this series and this book from beginning to end. Darwin students J Liley and E Charalampadou generously assisted with indexing. The annual lecture series was inspired by Andy Fabian more than thirty years ago and under his stewardship this and all preceding series have come together successfully as the annual flagship event of Darwin College.

1 Ebola

The Plague of 2014/2015

JONATHAN L. HEENEY

The outbreak of Ebola in West Africa was an unprecedented epidemic of a highly fatal infection. Unexpectedly, after 12 months it was still raging, taking more than two years to control. It threatened international borders, dominated headlines and evoked fear around the globe. A contagious viral disease that causes a fatal haemorrhagic fever, Ebola is publicly perceived as having the capacity to cause a 'Plague' of catastrophic consequences. Plagues historically have not discriminated; they have afflicted humans of all levels of society, having both immediate and long-term consequences. The aftermath has had significant influences on public health, medicine, education, communication, societal practices, national economies, as well as cultural and religious beliefs.

The 2014/2015 outbreak was the largest epidemic caused by a Filovirus such as Ebola, since *Zaire ebolavirus* was first identified in the 1970s. Having infected over 28,000 persons in West Africa it was slow to be recognised and contained. Unfortunately, resolving this epidemic was difficult and protracted, taking two years because of the unexpected appearance of late cases, some without links to someone with clinical disease. This created apprehension in the countries of Guinea, Sierra Leone and Liberia; nations which were economically and socially anxious that the epidemic was officially declared over.

This was a plague of modern times and revealed a number of unexpected outcomes. It changed paradigms not only in our understanding of this disease but also how new vaccines, medicines and diagnostic technologies could be fast-tracked, and changed how international responses to global infectious disease threats could be handled better in the future. This epidemic impacted on the small village communities as well as large urban centres. It changed practices and local perceptions of infectious diseases, re-established the critical importance of basic health infrastructure and highlighted the crippling impact that such gripping infectious diseases have on the struggling economies of developing countries.

Ebola: The Most Feared Contagion

In central Africa *Ebola* viruses are known to periodically strike remote villages, causing a terrifying haemorrhagic fever with high mortality. Unexpectedly in December 2013, at a distance of 3000 km west of the known heartland of Ebola in Central Africa, an outbreak of a mysterious disease began to spread, extracting a terrifying toll and killing many. Initially diagnosed as cholera, it took until March 2014 to confirm that the disease was actually caused by an Ebola virus (EBOV). By then, unlike other Ebola outbreaks before, it had spread to populated urban centres and neighbouring countries.

EBOV is the only member of the species of *Zaire ebolavirus*, within a family of RNA viruses known as Filoviruses, so named because of their unusual intertwined filamentous shape (Figure 1.1). Members of this family of viruses include Ebola's African cousin in the east, Marburg virus, which resides in the Egyptian fruit bat.[1] The genus *ebolavirus* consists of at least four distinct species of African haemorrhagic fever viruses (Zaire, Reston, Sudan, Taï Forest and Bundibugyo viruses), and a fifth, an unusual Asian relative called Reston virus. It is found in pigs in the Philippines and causes haemorrhagic fever disease when it spills over to monkeys.[2] To date no cases of severe haemorrhagic fever caused by Reston virus have been reported in humans despite documented exposure.

Small and ultimately contained outbreaks of Ebola virus haemorrhagic fever have been reported in Central Africa ever since the virus was discovered by Peter Piot and co-workers in 1976. Mortality rates in EBOV outbreaks have been reported to be as high as 90 per cent. Until 2014 loss of life from a single outbreak had claimed no more than 400 lives.[3] Over twenty outbreaks have been reported (Figure 1.2) since 1976 in the Central African countries of the Democratic Republic of

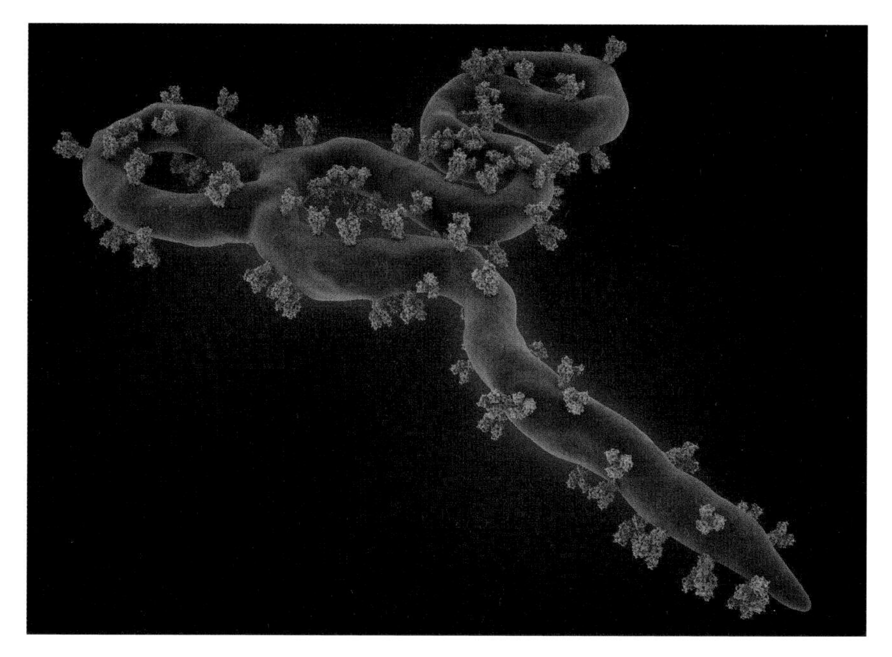

FIGURE 1.1 The Ebola virus, a Filovirus, is one of the causes of haemorrhagic fever in humans and non-human primates. It contains a single-stranded negative sense RNA genome and is approximately 19 kbases long with intact virions being 80 nm in diameter and up to 14,000 nm in length (image courtesy of the Wellcome collection). Although it is slightly narrower than the spherical HIV virus at 100 nm, its long filamentous shape makes it one of the largest RNA viruses (1 nm is one billionth of a meter).

Congo (DRC), the Congo, Uganda, Sudan, Gabon, South Africa and the Côte d'Ivoire, with the DRC having the most cases.

In general, Ebola-related disease includes haemorrhagic fevers which begin 2–21 days after exposure to an infected animal or person with the development of generalised symptoms including fever, chills, depression and muscle pain, usually followed by nausea, vomiting, diarrhoea, chest pains and bleeding from mucosal sites. After 6–16 days of severe disease many patients die of shock and multi-organ failure.

In the summer of 2014 there were two separate outbreaks of Ebola virus disease(EVD) ongoing. One in the DRC which was relatively contained, and the larger ongoing West African epidemic that was raging out of control. Interestingly, the first case in that DRC outbreak began

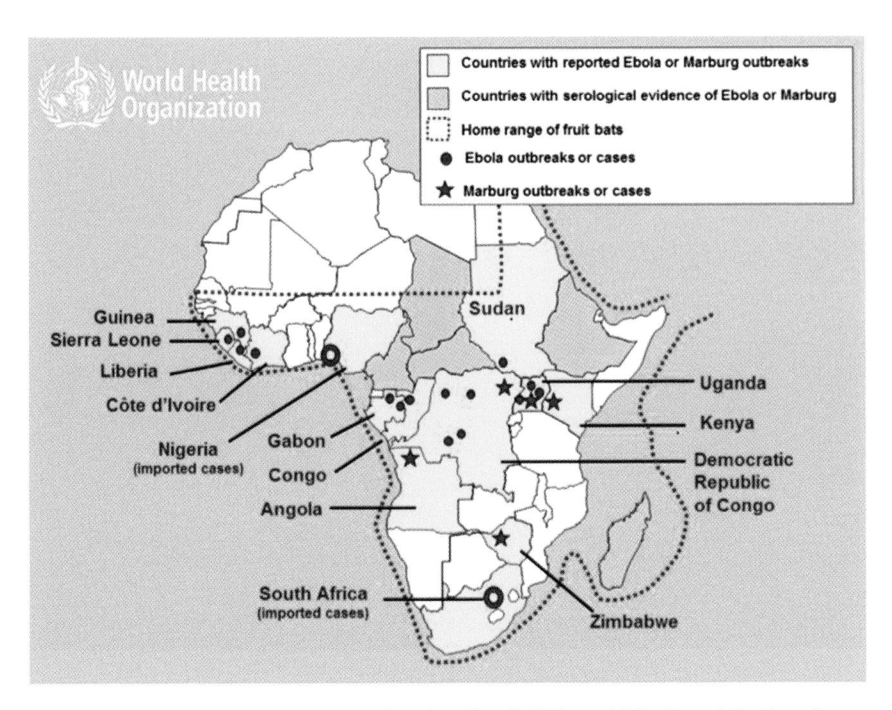

FIGURE 1.2 Location of outbreaks of Ebola and Marburg infections in humans and the wide-range distribution of fruit bats in Africa. Prior to the 2014 West African outbreak, significant human epidemics had not been found west of Gabon (image courtesy of the WHO Geneva).

with a case of a pregnant woman who had prepared and eaten meat from a monkey her husband had brought home. Unfortunately, the local doctor and three health workers who performed a caesarean section on her also became infected and died.[3] Ebola outbreaks most often begin with humans consuming bushmeat, one of the major risk factors for contracting Ebola in endemic regions and, again, this practice initiated the outbreak of Ebola in the DRC (the seventh known since Ebola was first identified). Unfortunately, health care workers are also at very high risk, especially early in outbreaks before a diagnosis is reached and the danger is recognised and Public Health authorities are informed. Because regional knowledge from previous outbreaks exists in the DRC there is awareness and existing strong coordinated outbreak response procedures in place. The 2014 DRC epidemic was quickly recognised, contained and limited to sixty-six fatalities. Within three months that outbreak was declared over. Why, in contrast, was the West African Ebola epidemic so large and why did it take so long to bring under control?

Ebola as a Plague

Why did the Ebola outbreak in West Africa reach plague-like dimensions? What failed given today's knowledge of modern medicine, the advanced disciplines of global and public health, outbreak control protocols of quarantine, hygiene and acute care medicine? In the right circumstances when today's medical technologies are combined with well-trained global response teams, they should be able to prevent most disease outbreaks from becoming plagues of biblical proportions or as severe as the Black Death of the middle ages. With the exception of a novel strain of highly pathogenic Avian influenza, even contagious airborne viruses such as SARS and MERS can be controlled if effective public health measures are rapidly enforced and the outbreaks effectively contained in countries with good established health systems. Fundamentally, it came down to the lack of knowledge and resources at the national and local levels. There was no prior knowledge of Ebola, inadequate health care infrastructure, no preparedness or capacity to identify the disease, or to respond, which made the poor developing countries of West Africa extremely vulnerable.

The regional experience of Ebola outbreaks in the DRC explains in part why outbreaks in Central Africa are capable of being contained within months and why in West Africa this new infection reached modern plague-like proportions devastating three geographically adjacent and culturally intermixed nations. One very significant factor is their regional history. Political turmoil and violent civil wars left weak struggling governments with bankrupt economies and precarious infrastructures. Neighbouring Côte d'Ivoire, Liberia, Sierra Leone and Guinea have a combined population of approximately 35 million with diverse cultural, ethnic and religious backgrounds that migrate almost freely across very porous and artificial colonial borders. In the preceding decades the devastating civil wars in Liberia spilled over to Sierra Leone. The large common borders with Guinea resulted in it being the recipient of those either seeking asylum or retribution. Similarly, during the bitter civil war in Côte d'Ivoire, Guinea was the largest recipient of both refugees and combatants, an unfortunate legacy from its three warring neighbours. Poor and traumatised by war and political instability, what health care and related infrastructures that did exist were stretched, in disarray and in tatters. Not surprisingly, there was considerable mistrust in government institutions following the long period of war and political turmoil. On the eve of this outbreak the main concern of these nations was economic recovery, peace and democracy.

Ebola was unknown to West Africa and had not previously been reported in the countries of Guinea, Sierra Leone and Liberia. Their populations and governments were caught off guard without any historical experience. This disease was a surprising new entity. Unsuspecting and ill prepared, there was a critical delay in response. Compounding this, the disease symptoms overlapped those of malaria, typhoid fever, Lassa fever, cholera and others. Delayed and side tracked by a misleading diagnosis of cholera, (a cause of deaths in recent years in Guinea) identification of the true cause of the disease took far too long. With months of delay the infection spread rapidly and case numbers accumulated significantly. Finally, in March 2015 when test results requested by Médecins Sans Frontières (MSF) came back positive for EBOV, international teams and equipment for diagnostic units and high containment field hospitals were mobilised and arrived in West Africa within weeks.

Efforts to contain the outbreak were thwarted by the delayed diagnosis, which postponed the arrival of international outbreak teams. Their efforts to contain the epidemic were logistically challenged by poor infrastructure. The infection had quickly spread geographically across national borders by a network of poor roads through many villages to larger urban centres (Figure 1.3). In contrast to the remote villages of the DRC, where outbreaks have occurred and ultimately been contained, the West African Villages in Forest Guinea were relatively close and linked by a network of roads in border regions where tribal family contacts and cross-border visits were frequent. Most villages had several mobile telephones that were used to keep contact with relatives in other villages,

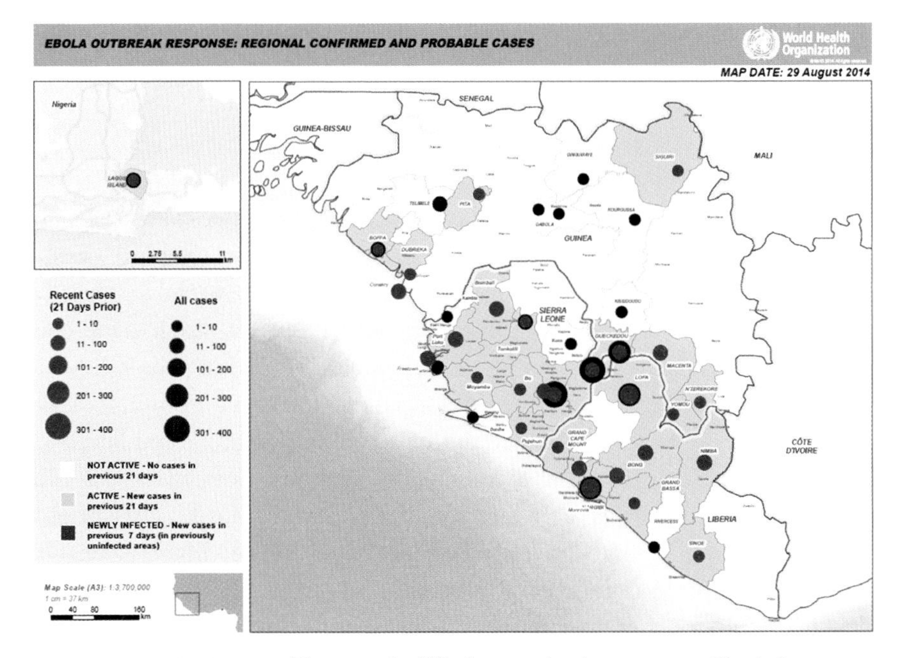

FIGURE 1.3 The spread of Ebola cases by August 2014. The index case was a two-year-old boy in the village of Meliandou, Gueckedou prefecture, Guinea. Next to die was his three-year-old sister, his mother and grandmother. The village midwife was treated and died in Gueckedou hospital and an unsuspecting health care worker from that hospital transmitted the infection to Macenta, and from there it spread to Kissidougou in February 2014. The rapid movement of infected people across the border with Sierra Leone and Liberia fuelled the epidemic. By August 2014 the epidemic had reached and established itself in the densely populated capital cities of all three countries (WHO outbreak map 29 August 2014. Courtesy WHO Geneva).

often across borders, and to make arrangements for sick family members to be transported by motorcycle chauffeurs to traditional healers and unsuspecting ill-equipped health care workers.

Cultures and Religious Beliefs

A number of societal factors contributed to the continued dissemination of EBOV infection even after the identification and arrival of international teams. Important amongst these factors were the tribal cultures, no prior knowledge or experience with Ebola, government mistrust and community customs and practices. Tribal regions extended past the artificial colonial boundaries which separated Liberia, Guinea and Sierra Leone. This region is ethnically very diverse with close family and community bonds. Visits of respect to funerals, including cultural customs such as physical contact with the deceased, played an important role in transmission. Mortuary practices include touching or sleeping beside the bodies of the deceased, and even the use of body parts to identify sorcerers, including secret exhumations and reburials. Each tribe and village has similar but often variable practices. The Kissi people have a custom of taking care of sick family and village members, even if necessitating travel to local hospitals to do their cooking cleaning and laundry. The hospitals recognise and accommodate such practices, even visits to the very ill to pay respects. If a person dies within the hospital, traditionally it is important that the body be repatriated to the village to enable the family to perform mortuary tributes.

The rapid arrival of international teams sent to curb the epidemic failed to recognise the importance of these customs, and they were often resisted and met with contempt when strict military-style quarantine practices were enforced. In a number of instances anthropological insensitivities led to violence by villagers, and in some cases mistrust of foreign outbreak response staff turned violent. Critically, the highly contagious nature of a corpse which had died due to Ebola was not compatible with mortuary practices of different villages. Ultimately, with the help of experienced anthropologists, education about the contagious nature of the corpse and dialogue with the village elders often resulted in a compromise, with the corpse being sprayed with disinfectant and delivered back to the village in a closed body bag, with the appropriate discussion about the contagious risk and care needed to bury the family member safely. Such accommodations eventually worked but they took time with careful dialogue and sensitive negotiations. Importantly, not all villages and villagers were co-operative. Early in the epidemic there was a great sense of mistrust and complot theories abounded.

Other important but subtle considerations included the important concept that different generations are buried separately. This necessitated the separation of the foetus from a dead pregnant mother so that each could be buried separately. Death in the time of Ebola meant that such practices would have life-threatening consequences for the living. Some villages understood and made compromises in the name of the living, but the fundamental respect for the future of the deceased, their relatives and the impact on the wider village community were issues that provoked violent responses if not handled with compassion and sensitive negotiated solutions. Violent responses by some village communities occasionally resulted in the unfortunate loss of life of some foreign Ebola response teams before these sensitivities were recognised and responsibly dealt with.

Interestingly, the experiences and actions taken by different villages early during the outbreak provided insight into the remarkable differences between village communities, some within 10 km of each other. On my second trip to West Africa, I visited villages in the Forest region of Guinea, many of which were the first to be hit by Ebola early in the epidemic, including Meliandou, the village which had the very first cases, a two-year-old boy and his family in early December 2013. To further understand the risk factors that played a role in the acquisition and spread of Ebola, in discussion with local authorities, we carefully selected villages to visit that had no cases of Ebola as well as neighbouring villages which did (Figure 1.4).

Our discussions with village elders and the local communities often revealed very different attitudes and experiences that had a major impact on Ebola spreading (or not) to their village. One village in particular was quite distrustful of outside influences, including the local and national governments, and white-western motives (who came, they said, only interested in protecting themselves). They rejected medical interventions, believing that the first cases of Ebola they experienced had followed recent vaccinations. They had since largely refused vaccines offered to the children of the village and treated government health care workers with considerable mistrust. In stark contrast, a neighbouring village 10 km down the road had experienced no cases of Ebola because of their own implementation of strict village quarantine practice as a result of a visit and discussion by an Ebola response team shortly after EBOV was determined to be the cause of the outbreak. This experience demonstrated that when approached properly with a dialogue that took into account

FIGURE 1.4 Village visits and community discussions in Forest Guinea. We visited, presented information and interviewed the inhabitants of villages of Gueckedou, Macenta and Kissidougou prefectures to discuss their health before and after Ebola came, if they hunted or ate bushmeat. We met with Ebola survivors and talked about their health since their recovery from Ebola haemorrhagic fever (Photo of villagers of Wondero with the author and members of the field team).

community concerns and cultural practices, outbreak awareness and quarantine practices work if response measures can be aligned successfully at the community level.

The implementation of strict hygiene practices, with well-recognised hand-wash stations, the difficult and sensitive modification of burial practices and avoidance of contact with persons and places with symptoms or cases, began late but eventually proved effective. My experience was that the people in the smaller village communities we visited were open, interested and engaged. In general, they were very willing and interested to understand diseases such as Ebola and wanted the assistance of our team to help improve the level of their health care. Although I only had the opportunity on that trip to visit just over a dozen villages, the cultural diversity of this sample was remarkable, having experienced five different indigenous languages, several different religions and various cultural practices.

Ebola's Origins and the Environment

In general terms it is estimated that 75 per cent of human infectious diseases have their origins from animals. The locations of animals reported to have evidence of Ebola virus infection (antibodies or EBOV genomic fragments) for the most part correspond to countries having Ebola outbreaks in humans, with the exception of seropositive bats (having antibodies to the virus) being identified in southern Ghana (comprehensively reviewed by Pigott et al. 2014[4]). Most cases reported were in gorillas, then chimpanzees, with only two reports of forest antelopes infected by Ebola. Non-terrestrial mammals suspected but not yet proven as reservoirs include three fruit bat species (hammer headed fruit bats, straw coloured and little collared fruit bats), based on indirect evidence. The undisputed evidence for the reservoir of Ebola's viral cousin Marburg virus has come from its direct isolation from the Egyptian fruit bat. Fruit bats remain high on the suspect list as they carry a number of other highly fatal human pathogens, including Lyssaviruses (Rabies-like viruses) and the Henipaviruses, Hendra and Nipha viruses (transmitted to humans by spill-over from bats to horses [Hendra] or pigs [Nipha] in Australia and Malaysia, respectively).

The tropical rainforests of Africa are amongst the most biologically diverse in the world, rich in animal species, providing the complex ecosystems for all of the Ebola reservoir species. One hypothesis I favour includes the seasonal congregation of the potential reservoir species during fruit-bearing season with the transmission of virus from the faeces or urine from fruit bat species above with spill-over to non-human primates and antelopes below. Bushmeat consumption is a major source of protein in Central and West Africa, and all potential Ebola virus carriers are sold and consumed. Furthermore, exposure to the ecosystems in which Ebola virus is thought to exist is increasing. Human populations in Africa are rapidly growing, and human activity – logging, farming and hunting – is encroaching into these ecosystems, increasing the risk of human transmission. The Guinean forests are no exception. Amongst the most biologically rich and diverse in Africa, these forests are also being heavily exploited for lumber, firewood, sources of food and potential agricultural land. The distribution of the West African chimpanzee is under threat. The local bushmeat hunters we interviewed in Macenta, Guinea, commented that sightings of chimpanzees were extremely rare compared to reports from the older generations. They concluded themselves that this was because the forests were receding.

Although the consumption of gorilla or chimpanzee meat has been associated with many of the outbreaks in central Africa, they are not thought to be natural reservoirs of EBOV per se (Figure 1.5). Large outbreaks of fatal disease associated with Ebola have been reported to cause devastating losses to wild great ape populations,[5] revealing that they are as susceptible, or even more so, to the pathology caused by Ebola. In contrast, observations of bats suggest that they may be the natural reservoirs in which infectious virus circulates without causing the large population die-offs reported in human and non-human primates. In this regard non-human primates can be considered as amplifying hosts, whereby the virus spill-over from bats rapidly replicates to high titres, weakening or killing large numbers of primates and possibly other species such as duikers (forest antelopes), making them easy bushmeat targets. This does not exclude bat to human transmissions, as bats are also consumed, but their capture and consumption is more rare and the exposed viral dose is potentially much lower (Figure 1.5).

A significant outbreak of Ebola-like disease in western chimpanzees occurred next door to Guinea in the Tai forest of the Côte d'Ivoire in 1994. Amongst the dead animals, a novel EBOV was identified and designated Tai Forest Ebola virus. A Veterinary pathologist performing an autopsy on one of the animals developed what was reported to be a Dengue-like febrile disease and was airlifted to Switzerland where she recovered.[3] No other human cases were identified. Until 2013, this was the furthest west an Ebola-related virus had been reported to naturally occur. Interestingly, despite its geographic proximity in the Côte

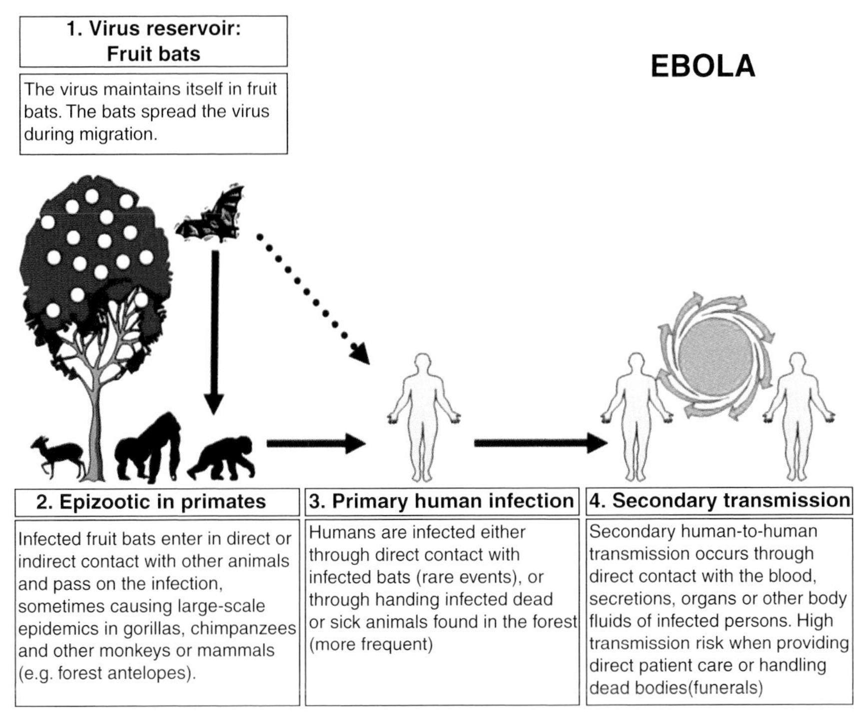

FIGURE 1.5 The Zoonotic origins of Ebola virus. The leading hypothesis is that EBOV naturally circulates in as many as three different fruit bat species that act as the natural reservoir of the virus. Seasonal and environmental effects may bring fruit bats together with other fruit-eating terrestrial species such as primates. However, these are not naturally resistant to disease and become infected to high levels. This effectively amplifies the amount of EBOV and increases the contagious risk of spill-over to humans. Some primates, especially gorillas and chimpanzees, are, like their human relatives, highly susceptible to Ebola virus disease and replicate the virus to high levels. This increases contagiousness of bushmeat, causing local outbreaks upon simple contact with body fluids. (image courtesy of the WHO, Geneva).

d'Ivoire, the Tai forest Ebola virus was not closely related to the variant of the Ebola virus that caused the human outbreak in neighbouring Guinea.

The first known case of Ebola developed in a two-year-old boy in the village of Meliandou, just a short distance from the city of Gueckedou. His three-year-old sister, mother and grandmother all died of haemorrhagic fever.[6] How the young boy acquired the infection, and if he was indeed the index case who acquired the infection from a reservoir animal, remains unknown. Indeed theories about the animal species that transmitted the infection to the young boy and his family are speculative. Similarly, it is not known how the Zaire strain of Ebola virus arrived in West Africa 3000 km away. Several theories exist, and the migration of infected bats is the leading suspect for transmission across such a distance. Using molecular clock predictions based on the rate of evolution of the Ebola virus genomes to compare human isolates in West Africa to those from Central Africa, data suggest that the Western and Central African EBOV diverged from each other approximately ten years ago[7].

Though it will be difficult to prove, a comprehensive survey of suspect species in Gueckedou and key regions of Forest Guinea may lead to clues as to which animal reservoirs may carry Ebola virus. It is unlikely, however, that we could ever conclude the event that led to the animal to human spill-over. Speculation that it may have been bats is based on the observation of a burnt-out hollow tree close to the village of Meliandou that contained bat faeces; however, no evidence of virus or even bats with antibodies was found. Molecular data derived presumably from bat faeces in the tree in Meliandou suggested that the tree had been inhabited by insectivorous bats, not fruit bats, which are the prime suspect EBOV reservoirs. The emergence and prediction of infectious diseases capable of causing human epidemics and pandemics remains an inexact science due to the number of fluctuating variables (climatic, ecological, genomic, zoologic) and their complex dynamics.

When nature's finely balanced ecosystem relationships change, such as a climatic event or change in food availability due to fires or deforestation, species may migrate or shift migration patterns to follow food sources. Epidemics or epizootics (an epidemic in an animal population) of infectious diseases that jump species require certain conditions to come together. Given what is known about Marburg and Ebola,[4] scientists have plotted the environmental and ecologic conditions needed to sustain the wildlife populations in which Ebola circulates and their geographic environment. This, combined with human population maps, has helped predict the larger regions where an Ebola outbreak may occur in humans. Not surprisingly, the interdigitation of the human population in the forested regions of West Africa, the natural flora and fauna with the suspect animal reservoirs, suggested it was ripe for an outbreak of Ebola.[8] Similar predictions for zoonotic infections can be combined with climatic events such as years of heavy rainfall and changes in vegetation as observed with Rift Valley fever outbreaks.[9] Viruses that are transmitted by insect vectors are influenced by shifts of the right insect species to a new geographical region, such as by global transportation (West Nile Virus in the Americas) or climate change. A change in the geographic distribution of the insect vector capable of transmitting certain viruses may result in the shift of diseases to human populations which have little or no immunity (recent examples include Dengue, Zika and Chickungunya viruses).[10] Such changes in animal reservoirs (and/or the vectors that may transmit them) may have caused the emergence of disease not previously observed in a geographic region and its human population.

Fortunately, not all viruses of animals are equally infectious to humans and many cannot be easily transmitted from one human to another. Such is the case of rabies infection, where human to human transmission does not normally occur. Other pathogens such as avian influenza viruses need to undergo a series of mutational changes before they can be readily transmitted to humans from poultry or pigs, and further mutations before they can become transmitted from humans to humans efficiently, and even more before they become re-assorted with influenza serotypes that are unknown to our immune systems and become potential pandemic pathogens. HIV is an interesting case in point. There are type 1 and type 2 human immunodeficiency viruses; HIV-1 became a globally distributed pathogen that is silently transmitted decades before the onset of AIDS in infected people. This virus had its origins in monkeys, but it required many infections in chimpanzees before it recombined with another AIDS-like virus also acquired by chimpanzees to become an infection that was easily transmitted from human to human and to become a global pathogen. Its cousin HIV-2 from monkeys causes a less aggressive disease, and has remained largely restricted to West Africa (for further reading see Heeney et al. 2006[11]). Unlike HIV, EBOV causes an acute disease within 2–21 days of contact with body fluids from someone who has died or is clinically ill with Ebola viral disease. As mentioned earlier, some outbreaks of EBOV have had high death rates. This 2014/2015 West African outbreak however differed, having mortality rates below 50 per cent.

Humans: An Expanded Reservoir of Ebola?

The West African Ebola plague introduced a surprising new dimension. It was generally thought that if Ebola patients survived the acute haemorrhagic crisis, their immune systems would eliminate any residual virus. Before Ebola survivors were released from Ebola treatment centres (ETCs), protocols usually required patients that have returned to good health to have at least two negative blood tests for the genome of the virus (called a PCR test) at a 2-day or more interval. However, compared to previous Ebola outbreaks the population of Ebola disease survivors in this epidemic was extremely large. Of the 28,000 people infected, it is estimated that approximately 16,000 survived EVD. Accumulating evidence with these large numbers has revealed that despite being declared negative from blood samples, in some patients the virus may persist in deeper compartments within the body where the immune system is unable to clear the virus. This includes the eyes, brain, testes and joints where EBOV has been found. In 30 per cent of male Ebola survivors, virus genomes can be detected in semen 6 months or longer after they have recovered from the disease. Sexual transmission from male survivors has been documented and female to male transmission has been suspected. Breast milk transmission has also been observed, indicating that mother to child transmission may also occur from Ebola female survivors. Importantly, while an Ebola survivor is healthy, the risk that they may secrete the virus is low and the risk of transmitting the virus by casual contact is negligible. However, the risk from intimate sexual contact, or mother to child transmission during or following childbirth are concerns. A major dilemma is how to follow up these individuals without stigmatising them, considering that they already have survived the traumatic life-threatening experience of Ebola disease. Continued community support and acceptance of this large population of Ebola survivors is crucial for their well-being, while responsibly ensuring that public health precautions such as the use of condoms is encouraged.

During a visit to one village in Macenta prefecture, I met a remarkable woman who reported that Ebola came to her village after someone had returned from a funeral in Liberia. Both she and her father developed symptoms and were treated by the ETC in Gueckedou, where they fortunately survived. Having listened to the health care workers who came to their village to describe Ebola symptoms and prevention measures, when she developed symptoms, rather than requesting assistance and possibly exposing someone else from the village, she walked a great distance alone to the ETC and submitted herself for quarantine and care. Fortunately, both she and her father survived EVD and after observation and negative viral tests, were allowed to return to their village. Remarkably, she was energetic and eager to return to the fields despite her near death experience with Ebola.

There was a growing consensus that follow-up medical monitoring of Ebola survivors was needed for two reasons: (1) to identify and to provide treatment for those individuals with long-term inflammatory conditions as a consequence of Ebola infection and (2) to provide counselling to survivors who are at risk of carrying the virus, and to advise them to take measures to prevent them from transmitting their infection, such as the use of condoms. While the incidence of transmission from a survivor is statistically low, it would only take one case for the epidemic to flare up again. The World Health Organization (WHO) declares a country free of Ebola if there have been no clinical cases 42 days after the last case has been demonstrated to be negative. The 42-day rule is based on doubling the 21-day interval, the longest period of time it may take someone to develop clinical disease following exposure to someone with Ebola disease. The 42-day countdown begins after the last known case has died or been tested negative twice at a 2-day interval. The test measures the presence of virus genomes in blood samples, so a negative test indicates that the body has cleared the infection from the body's circulation, and thus has a very low risk of transmission. Unfortunately, there are no clinically acceptable tests to determine if Ebola survivors have cleared the virus from sites in the body that the immune system cannot easily access, such as the eyes, the brain, reproductive tract and joints.

Humans become infected when they consume bushmeat carrying Ebola virus. Once humans become infected and become ill (Figure 1.6, large

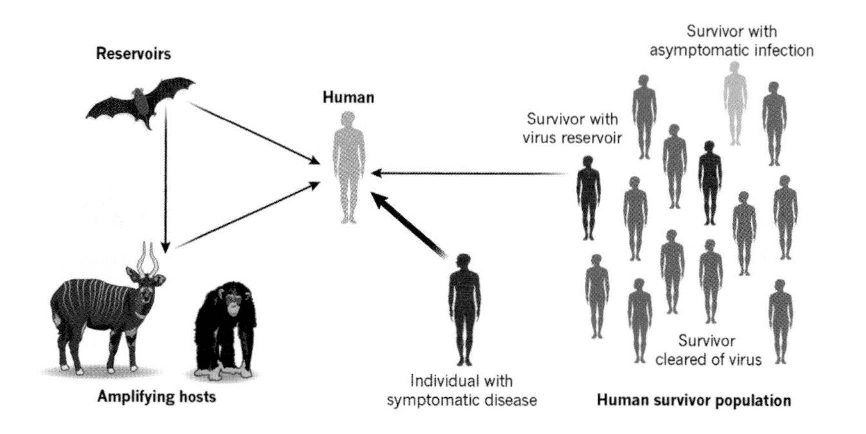

FIGURE 1.6 Ebola survivors and the human reservoir. Although not yet proven, current evidence suggests that the classical animal reservoir of Ebola virus, as for its cousin Marburg virus, are fruit bats. Three species of fruit bats may co-circulate Ebola virus and this may require naïve juveniles or naïve colonies of bats to sustain circulating virus in bat populations, moving as they overlap and migrate. Animals (such as great apes and forest antelopes) that occasionally/seasonally may share the fruit bat's habitat when certain fruit-trees have ripened fruit are exposed to bat urine, faeces and saliva when foraging for fruit in season.

figure shown in red), they become the primary source of viral infections for other humans and the human epidemic continues until the transmission chain is broken. However, some humans that become ill and survive may not completely clear the virus infection (Figure 1.6, small figure depicted as red) with virus residing in their central nervous system, eyes, joints or reproductive tracts. Pregnancy is a factor that may allow a carrier state to persist, with increased risk of transmission at or following childbirth. If immunity wanes and the virus replicates again to high levels EBOV may be transmitted from these previously quiescent infections. Thus some survivors may be transmitters long after quarantine periods have ended. Interestingly, some exposed individuals may be resistant to clinical disease but carry the virus (Figure 1.6, asymptomatic survivor shown in light grey above) (Image courtesy of *Nature*, in Heeney, 2015[12]).

The late flare-up of Ebola in Liberia is a case in point. After first being declared free by the WHO in May 2015, Liberia had a cluster of new cases in late June. Again after a 42-day period of no new cases the WHO declared Liberia free for a second time in September 2015. However, on 24 November 2015, a fifteen-year-old boy died of Ebola disease about 20 days after his father had developed Ebola. His eight-year-old brother developed the disease as well, but like his father was successfully treated and survived. All evidence suggests that the family acquired the infection from the mother, who had given birth to a new baby boy in September of 2015. The baby did not develop Ebola disease but the mother had been admitted to hospital a month after giving birth with fever. Both she and her baby were found to have antibodies to EBOV which may have contributed to protecting them from Ebola disease. Investigations revealed that more than a year earlier in July of 2014 she had cared for her brother who shortly after died of Ebola. She had become unwell at the time and her pregnancy at that time concluded in miscarriage in August 2014. This suggests that she survived as a relatively asymptomatic survivor (Figure 1.6), but upon another pregnancy and successful childbirth in 2015 she unknowingly transmitted the infection to her husband sometime after childbirth. This is not the only example of a survivor transmitting the virus. Male to female and female to male transmissions have also been reported, as has a case of breast-milk transmission.[12] This case, however, revealed her relatively asymptomatic brush with Ebola in 2014 that went undetected. This was followed with mild recrudescence a year later after another pregnancy that tragically led to three male members of her family getting Ebola disease, with one family member dying.

Another question concerns Ebola disease resistance in individuals who were highly exposed to the virus. During my visits to villages affected with Ebola, I met a remarkable number of individuals who were with, or cared for, infected family or members of the community who had the disease, but who did not develop Ebola themselves. One village doctor told me of caring for the very first cases of Ebola in his community before they knew what the disease was. He was personally administering fluid therapy and caring for patients with his bare hands. Despite high-level exposure and no protective equipment, he did not develop the disease. It remains to be seen if this was due to an asymptomatic infection, or if he was genetically resistant to the viral infection itself, such as a genetic-like resistance as observed in the HIV/AIDS epidemic.[11]

There is clearly an important need for delicate handling of the Ebola survivor question to avoid the risk of Ebola survivors being ostracised like lepers from their communities and forced into denial or hiding. Medical assistance and follow-up needs to be offered to survivors and balanced public health advice given without raising public fear and negative attitudes in communities (Figure 1.7). Tragically, the need for medical follow-up of Ebola survivors was recently illustrated by the case of a British nurse who had become infected by Ebola virus while caring for patients in West Africa. She developed the Ebola disease but survived due to heroic and expert care in the UK. However, 9 months later she became ill again with an unusual neurological meningitis-like presentation. Fortunately, she was quickly hospitalised and recovered following experimental anti-viral treatment. Her case emphasised the risk of relapse and secondary disease in some Ebola survivors.

FIGURE 1.7 Ebola survivors. There are an estimated 16,000 or more people that have survived Ebola virus exposure and Ebola disease in West Africa. Despite early reluctance, the majority of individuals have been welcomed back to their families and communities. There are important concerns, however, for their community support. Despite being negative for the virus in their blood, a number of Ebola survivors have virus that persists within their bodies which may flare up and cause a different type of disease. While most are thought to be otherwise healthy and have cleared the infection, an unknown minority of survivors may carry virus in their eyes, joints, reproductive tract and central nervous systems, with in some cases disabling arthritis, ocular disease, hearing disorders and even encephalitis.

This hidden reservoir of Ebola has now become more apparent and more of a public health issue because of the large number of Ebola survivors living throughout West Africa (Figure 1.7). The development of diagnostic testing to confirm that Ebola survivors had successfully cleared the virus or were carriers would greatly aid the management and, potentially, the treatment of the Ebola survivors at risk.

Ebola: The Game Changer for Human Clinical Trials

The unprecedented scale of this Ebola epidemic with its relatively high mortality rate motivated a major shift in the cautious pragmatic clinical trial process used to evaluate experimental drugs and vaccines. Relatively early in the epidemic, the large number of fatalities, not only amongst the affected population but also the heroic local and international health care workers in the ETCs, was traumatic. The experimental drug ZMapp was one of the first to be offered in limited supply to a few who developed EVD. It is a cocktail of three monoclonal antibodies produced from tobacco plants which block virus entry, two from Canada's Public Health Agency in Winnipeg, and the third from the US Army Medical division. It was given to a few patients but supplies soon ran out in August 2014. Another drug, GS-5734, an inhibitor of viral replication, was fast-tracked in the spring of 2014 and was used in a small number of patients (including the UK nurse with meningitis-like relapse) who were critically ill and had no other treatment options. The urgent need during the epidemic did not allow for the pragmatic planning and safeguards that are built into stepwise clinical trials that put safety first. Early results suggested that of the six or so candidate anti-viral drugs, ZMapp and Favipiravir showed early evidence of benefit. However, they were often used on patients with advanced disease but with insufficient numbers to be conclusive. Complicating the desire to fast-track new drugs was the established processes and study design debate. Unfortunately, as companies ramped up production of candidate anti-Ebola drugs for larger scale clinical trials, the number of cases began to decline, too few too late to make a difference for this outbreak.

The experience with experimental vaccines, however, proved somewhat more favourable. Several vaccine candidates were available in concept, but not at clinical (GMP) grade sufficient for immediate use in humans. Phase 1 safety trials in human volunteers were performed for a small number of vaccine candidates, as it was necessary to use these vaccines in humans who were not at risk of Ebola infection. One vaccine candidate, VSV-EBOV, was ready in time for Phase II and Phase III trials in Guinea (Figure 1.8). Early data appeared to show a positive effect of containing EBOV spread through an approach called 'ring-vaccination', a vaccine strategy used in Veterinary medicine to contain an outbreak by vaccinating all potential contacts around known cases. The first results

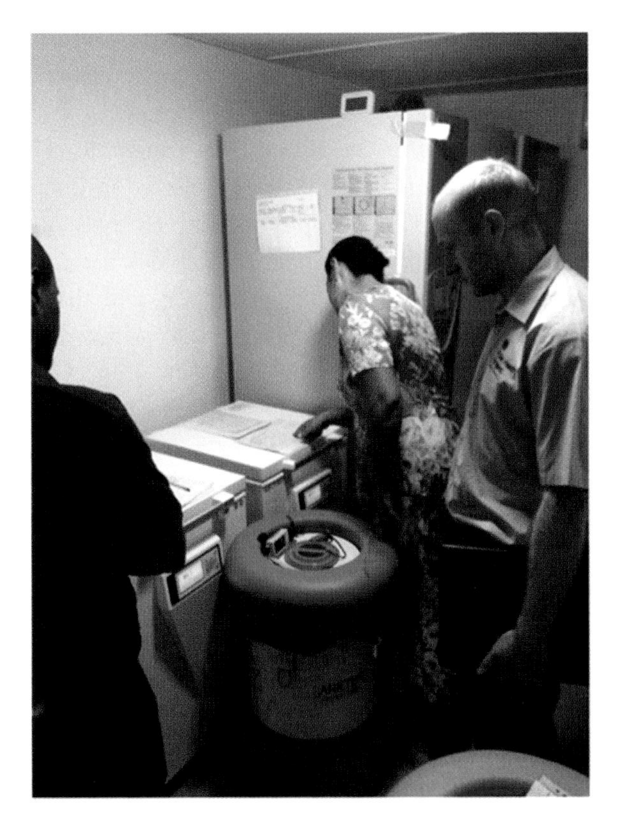

FIGURE 1.8 The Ebola rVSV-ZEBOV vaccine being prepared for distribution to sites in Guinea where people who were exposed to an individual with Ebola disease were vaccinated to halt the infection from spreading in the community.

suggest that this vaccine is probably efficacious at preventing disease. This vaccine candidate and strategy were deployed in the last stages of the epidemic for the containment around a flare-up case(s) and the prevention of EVD in people exposed to someone possibly infected with the virus.[13]

Ebola's plague-like Consequences

Beyond the loss of life and suffering, the consequences of the West African Ebola outbreak were far-reaching. This was a regional plague that threatened to spread internationally and, indeed, was exported by infected people travelling to North America, Europe and other countries in Africa. In contrast to historical plagues, there was a mobilised international effort that, although delayed, ultimately controlled and contained what centuries ago could have become a devastating global plague. Regionally, the three affected countries have suffered very significantly, both in terms of loss of life, economic–loss, as well as loss of infrastructural development. Many countries pulled their teams out of projects which were involved in building roads, hydro-electric stations and improving transportation infrastructure. Stigma and aversion reduced already low international investment. Strict controls at borders and restrictions on the movement of people slowed trade and business to low levels. The governments of affected countries focused most of their limited resources on the national disease control effort and many senior government workers were sent to their home provinces to aid in epidemic control measures. The existing health infrastructures that existed were strained under the sheer stress of contact tracing, outreach and communication and coordination with ETCs, WHO and international relief organisations such as MSF. Overall, the regions per capita income was expected to decline by $18.00 USD/year until 2017.[14] The United Nations Development group estimated that West Africa would lose an average of $3.6 billion USD per year between 2014 and 2017 due to decreased trade, border closures, cancelled flights and reduced foreign investment. The World Bank estimated that Guinea, Sierra Leone and Liberia together would lose $1.6 billion USD in lost economic growth in

2015 alone.[15] In addition, the outbreak impacted on future economic potential by causing some schools to close, reducing or restricting travel, while existing infrastructure slowed markedly. Consumer confidence in the region was eroded during the outbreak and a declaration of zero cases was an important milestone for economic recovery to begin. The outbreak had a general negative impact on the region, especially when Ebola virus transmissions spilt over to Mali, Senegal and Nigeria. However, the rapid containment of the disease and the follow-up declaration that these three countries were Ebola-free buoyed confidence.

Interestingly, the post-Ebola economic outlook may not be all gloom and doom. As Ian Morris pointed out in his paper on 'Plagues and Socioeconomic Collapse',[16] for these very poor African countries there may be a silver lining. The international response has brought awareness and change, and has focussed the minds of these countries on fundamental problems, ranging from health care, education and infrastructure. During my visit in November 2015 the government of Guinea announced that it was building a number of new hospitals. Major improvements in roads were already starting to happen. There was cautious optimism in Guinea with the recent election that the progressive plans by the newly re-elected government would continue. Perhaps in the long run the historical observations that calamities such as plagues may ultimately result in positive effects, benefit economic recovery and accelerate progress that otherwise would have moved on incrementally may prove correct.

Conclusions

The Ebola epidemic of 2014/2015 was a global threat because of international air travel, with infected cases reaching Spain, the USA, the UK, France, Germany, Norway, Switzerland, Nigeria and Italy. Most involved exposed and infected health care workers, missionaries or people visiting families in affected countries. These cases were for the most part rapidly contained, with new cases primarily affecting hospital staff where the infection was quickly identified, contained and controlled (but not without loss of life). Would the global threat of Ebola spreading have existed without air travel 100 or 200 years ago?

One hundred years ago the world was on the brink of WWI. There was no WHO until after WWII, but the War to end all wars involved a massive mobilisation of armies and populations, with the international transportation of troops by sea on a massive scale. With respect to West Africa the trans-Atlantic export of West African slaves to the new world had ended, but the African nations were under harsh and often cruel European colonial rule. Africa was pulled into WWI, and this pitted German-controlled territories primarily against Anglo- and French-ruled countries. In West Africa, Gambian, Ghanaian, Nigerian, Sierra Leonean and Beninese troops had invaded German Togoland, and although there were military transports to and from Europe information on troops and issues affecting troop strength were kept quiet. The concept of viruses was embryonic, and though the world was familiar with the diseases caused by plagues such as smallpox, the Black Death ('The Plague') and cholera, the agents that caused them were not well understood. The fundamental practice of quarantine was well-established and practised during the war. However, four years later a monstrous epidemic of highly pathogenic influenza defied conventional quarantine practices, causing an estimated 50 million deaths. The Spanish flu, as it became known, cost many more lives than WWI itself and is attributed by some to have directly or indirectly contributed to the end to WWI.

Two hundred years ago, the Napoleonic wars had recently concluded and British political desire to end the slave trade from the west coast of Africa was having an effect. Despite the abolition of slavery in the British empire in 1807, it was still going on in the British West Indies until 1833. Could Ebola have successfully made the trans-Atlantic voyage on a slave ship? Outbreaks of disease and loss of significant numbers of life were reported on slave ships, no doubt facilitated by the terrible conditions aboard them. Numbers of deaths were not only reported on board but also by the ports that shipped and received slaves from West Africa. Some ports receiving slaves in the New World enforced a 40-day quarantine for crew members as well as their human cargo. Trans-Atlantic voyage times from West Africa to the New World in the early 1800s were more than 40 days, almost twice the longest period of time that symptoms would appear in an EBOV-infected person, similar in duration to the 42-day case-free period the WHO now requires before a country can be declared Ebola-free. The 40-day sailing time combined with the additional 40-day quarantine period imposed at most importing ports in the new world would suggest that the transit time by ship plus the quarantine period upon arrival in the New World would have reduced the opportunity for Ebola to be exported to the New World. That said, without a continuous human chain of transmission from a person with Ebola disease symptoms, the epidemic would die out. The establishment of a human reservoir without modern acute disease treatment for Ebola would result in a much higher mortality rate and the disease would most likely be self-extinguishing in a confined, relatively small population.

We know that smallpox, measles and venereal diseases were successfully exported to the immunologically naïve indigenous populations of the New World, but they have considerably different disease dynamics. Fortunately, the transmission of Ebola requires direct contact with body fluids, or contact with a clinically ill or dead patient. Transmission of EBOV from persons who do not have fever is less likely and differs from Flu (influenza). A key difference is that Flu-infected persons are contagious up to 10 days before the onset of illness. In addition, Flu is far more contagious as it is transmitted by small aerosol droplets. Thus the quarantine practices that were developed in Italy in the Middle Ages (quaranta giorni = 40 days) for the control of the Black Death ('The Plague') would have been effective against Ebola if strictly adhered to. So why then did Ebola reach modern plague-like proportions in West Africa when outbreaks in the Central African DRC are most often contained to less than 100 cases? In summary, there are three major factors, 1) awareness, 2) preparedness and rapid response and 3) isolation and containment to small populations. In essence, the delay in recognition and lack of early containment led to dissemination to large population centres, including the capital cities of Monrovia, Freetown and Conakry. Essentially, these countries lacked the resources, experienced personnel and facilities to deal with this type of outbreak without international assistance. Other contributory factors precipitated the epidemic but many such as cultural beliefs, poverty and poor health care infrastructure are largely shared features between the DRC and the three affected West African nations. For the late international response the WHO have received much criticism and undergone major reforms in order to respond more quickly to infectious disease outbreaks.[17] Time will tell how the international community will respond to the next outbreak now that West Africa has been declared Ebola-free. One interesting statistic that haunts us is that within 24 months of an Ebola outbreak, there is a 50 per cent chance of another outbreak occurring in that country. The animal reservoir that carries the virus in West Africa, its distribution and the dynamics that impact on cross-species transmission remain unknown. Two years after the outbreak I personally observed bushmeat (including species known to carry EBOV) for sale along the roadside near Macenta in Guinea. Deforestation, primarily fuelled by logging, agriculture and mining, is continuing to cause changes in the Guinean forest which affect the balance of species, putting increased concentration on desirable food supplies. Such ecosystem changes influence the migration, behaviour and possibilities of chance interaction of reservoir species, amplifying host and human contact, thus increasing the opportunity of cross-species spill-over of EBOV. Many village communities have always relied on the forest for food and fuel while others have turned towards agriculture or grazing small numbers of livestock. For those that do not have alternative sources of food, they will continue to be dependent on what they can obtain from the forests for food and livelihood. It is unlikely that all village communities will be able to find sufficient amounts of protein from other sources than bushmeat in the near future.

This epidemic has had an impact on international response efforts. In addition to the changes in the WHO's policies and practices for outbreak control and response, significant advances in how clinical trials of new vaccines and therapeutics can be rapidly evaluated and deployed in an outbreak setting have been pioneered. One major success story is the identification of a new vaccine candidate for the control of the spread of infection and prevention of Ebola viral disease. Rolled out in a unique ring vaccine trial format in early 2015, the first positive results were reported by the WHO team in July of the same year (see Figure 1.8).[13] Not only has this vaccine been used to ring vaccinate around flare-up cases in the last 6 months of the outbreak in an effort to bring the epidemic to a conclusion, it has provided a pathway and model for which future outbreak responses can be deployed with experimental vaccines.

The economies of the three most severely affected nations are likely to slowly recover after being declared Ebola free. Paradoxically, there is an opportunity that the health care systems may potentially benefit from this epidemic. This opportunity to learn from the 2014 Ebola outbreak that plagued Western Africa will be dependent on investment. However, that is unlikely to come from the struggling national economies. Without international aid and investment, there remains the risk that the newly learned public health practices and health care advances made during the outbreak will revert to the old ways that existed before Ebola arrived. Continued international assistance in the short to medium term is not only necessary but key to sustaining this progress towards better health care.

Finally, the lasting legacy of this Ebola plague may not be fully revealed. We now know and are faced with the largest, unprecedented human reservoir of Ebola in the history of medicine. The optimistic outcome is that Ebola survivors will continue to recover and ultimately enjoy reasonably good health and support within their communities. More realistically, an unknown percentage of Ebola survivors will have persistent EBOV infection. Some may develop debilitating arthritis, eyesight and hearing impairment with the risk of viral meningoencephalitis. To what extent and how long these conditions may persist or possibly recur is currently unknown. Already discussed is the concern of late stage transmission and Ebola disease flare-ups with new human transmissions long after the epidemic has been declared over. For these reasons counselling and continued monitoring are needed, but it is highly unlikely that all of the large, scattered and occasionally unwilling population of survivors will have access to or desire for follow-up medical support even if it could be afforded and supplied. Will there be occasional major flare-ups as we have recently seen, or will the virus possibly mutate under long-term immunological pressure to change the spectrum of the disease it causes? There are also people in West Africa with HIV/AIDS that have become infected with EBOV. The persistence in persons with compromised immune systems with infections such as HIV can be expected to be problematic and cause post-Ebola disease. The HIV/AIDS epidemic has fuelled the emergence of antibiotic-resistant TB (tuberculosis) that in itself is becoming one of the single major global health problems facing mankind. Time will tell, but the Ebola plague of 2014/2015 will remain an epidemic and a lesson that we must neither forget nor walk away from.

References and Further Reading

1. Suzuki Y., Gojobori T. (1997) The origin and evolution of Ebola and Marburg viruses. *Mol Biol Evol.* Aug;14(8):800–6.
2. Geisbert T.W., Feldmann H. (2011) Recombinant vesicular stomatitis virus-based vaccines against Ebola and Marburg virus infections. *J Infect Dis.* Nov;204 Suppl 3:S1075–81. doi: 10.1093/infdis/jir349. Review.
3. Alexander K.A., Sanderson C.E., Marathe M., Lewis B.L., Rivers C.M., Shaman J., Drake J.M., Lofgren E., Dato V.M., Eisenberg M.C., Eubank S. (2015) What factors might have led to the emergence of Ebola in West Africa? *PLoS Negl Trop Dis.* Jun 4;9(6):e0003652. doi: 10.1371/journal.pntd.0003652. eCollection 2015. Review.
4. Pigott D.M., Golding N., Mylne A., Huang Z., Weiss D.J., Brady O.J., Kraemer M.U., Hay S.I. (2015) Mapping the zoonotic niche of Marburg virus disease in Africa. *Trans R Soc Trop Med Hyg.* Jun;109(6):366–78. doi: 10.1093/trstmh/trv024. Epub 2015 Mar 27. Review.
5. Bermejo M., Rodríguez-Teijeiro J.D., Illera G., Barroso A., Vilà C., Walsh P.D. (2006) Ebola outbreak killed 5000 gorillas. *Science* Dec 8;314(5805):1564.
6. Baize S., Pannetier D., Oestereich L., Rieger T., Koivogui L., Magassouba N., Soropogui B., Sow M.S., Keïta S., De Clerck H., Tiffany A., Dominguez G., Loua M., Traoré A., Kolié M., Malano E.R., Heleze E., Bocquin A., Mély S., Raoul H., Caro V., Cadar D., Gabriel M., Pahlmann M., Tappe D., Schmidt-Chanasit J., Impouma B., Diallo A.K., Formenty P., Van Herp M., Günther S. (2014) Emergence of Zaire Ebola virus disease in Guinea. *N Engl J Med.* Oct 9;371(15): 1418–25. doi: 10.1056/NEJMoa1404505. Epub 2014 Apr 16.
7. Gire S.K., Goba A., Andersen K.G., Sealfon R.S., Park D.J., Kanneh L., Jalloh S., Momoh M., Fullah M., Dudas G., Wohl S., Moses L.M., Yozwiak N.L., Winnicki S., Matranga C.B., Malboeuf C.M., Qu J., Gladden A.D., Schaffner S.F., Yang X., Jiang P.P., Nekoui M., Colubri A., Coomber M.R., Fonnie M., Moigboi A., Gbakie M., Kamara F.K., Tucker V., Konuwa E., Saffa S., Sellu J., Jalloh A.A.,

Kovoma A., Koninga J., Mustapha I., Kargbo K., Foday M., Yillah M., Kanneh F., Robert W., Massally J.L., Chapman S.B., Bochicchio J., Murphy C., Nusbaum C., Young S., Birren B.W., Grant D.S., Scheiffelin J.S., Lander E.S., Happi C., Gevao S.M., Gnirke A., Rambaut A., Garry R.F., Khan S.H., Sabeti P.C. (2014) Genomic surveillance elucidates Ebola virus origin and transmission during the 2014 outbreak. *Science* Sep 12;345(6202):1369–72. doi: 10.1126/science.1259657. Epub 2014 Aug 28.

8. Ebola Virus Disease and Forest Fragmentation in Africa. (2015) www.efasl.org/site/?p=1801.
9. Anyamba A., Chretien J.P., Small J., Tucker C.J., Formenty P.B., Richardson J.H., Britch S.C., Schnabel D.C., Erickson R.L., Linthicum K.J. (2009) Prediction of a Rift Valley fever outbreak. *Proc Natl Acad Sci USA.* Jan 20;106(3):955–9. doi: 10.1073/pnas.0806490106. Epub 2009 Jan 14.
10. Heeney J.L. (2006) Zoonotic viral diseases and the frontier of early diagnosis, control and prevention. *J Intern Med.* Nov;260(5):399–408. Review.
11. Heeney J.L., Dalgleish A.G., Weiss R.A. (2006) Origins of HIV and the evolution of resistance to AIDS. *Science* Jul 28;313(5786):462–6. Review.
12. Heeney J.L. (2015) Ebola: Hidden reservoirs. *Nature* Nov 26;527(7579):453–5. doi: 10.1038/527453a.
13. Henao-Restrepo A.M., Longini I.M., Egger M., Dean N.E., Edmunds W.J., Camacho A., Carroll M.W., Doumbia M., Draguez B., Duraffour S., Enwere G., Grais R., Gunther S., Hossmann S., Kondé M.K., Kone S., Kuisma E., Levine M.M., Mandal S., Norheim G., Riveros X., Soumah A., Trelle S., Vicari A.S., Watson C.H., Kéïta S., Kieny M.P., Røttingen J.A. (2015) Efficacy and effectiveness of an rVSV-vectored vaccine expressing Ebola surface glycoprotein: Interim results from the Guinea ring vaccination cluster-randomised trial. *Lancet* Aug 29;386(9996):857–66. doi: 10.1016.
14. UN News Centre: *Powerful effects of Ebola outbreak felt outside worst-affected countries, UN report finds.* www.un.org/apps/news/story.asp?NewsID=50313#.VnlZrFIT-Vo. Accessed March 10, 2015.
15. The World Bank: *Ebola: Most African Countries Avoid Major Economic Loss but Impact on Guinea, Liberia, Sierra Leone Remains Crippling* www.worldbank.org/en/news/press-release/2015/01/20/ebola-most-african-countries-avoid-major-economic-loss-but-impact-on-guinea-liberia-sierra-leone-remains-crippling. Accessed March 10, 2015.

16. Ian Morris 'Plagues and Socioeconomic Collapse', this volume, Chapter 6.
17. Advisory Group on Reform of WHO's Work in Outbreaks and Emergencies with Health and Humanitarian Consequences. www.who.int/about/who_reform/emergency-capacities/advisory-group/en/. Accessed 3 August 2015.

2 Plagues and History

From the Black Death to Alzheimer's Disease

CHRISTOPHER AND MARY DOBSON

A broad chronological overview of the plagues of the past and the present provides a basis for understanding how they have arisen, how they have affected societies over time, and how humanity has responded to the challenge of each new wave of deadly disease. Many puzzles still surround the plagues of antiquity, although recent techniques such as DNA analysis of skeletal remains are beginning to provide clues as to their causes; it has, for example, now been confirmed that the plague bacillus, *Yersinia pestis*, was responsible for the 'Black Death' of the mid-14th century. Many other infectious diseases or 'plagues', such as smallpox, typhus, cholera and influenza, have afflicted human populations over the centuries. As the causes of these diseases have become increasingly well understood, humanity has devised ever more effective means for their control. With the extension of human lifespan that has resulted from such progress, other medical conditions have become increasingly common, notably the neurological disorder, Alzheimer's disease. Although not infectious, this disease has become so prevalent in recent years that it is has been called a '21st century plague'. The intrinsic origins of this highly debilitating condition are now being explored intensively by scientists from many different disciplines leading, as with the plagues of the past, to new ideas as to potential strategies for its prevention and treatment.

Most people associate the word 'plague' with the 'Black Death' of the mid-fourteenth century, which in the space of a few years killed between a third and a half of the population of Europe, Asia and the Middle East – with frightening consequences. However, throughout history there have been many other widespread and devastating outbreaks of disease that have been labelled as plagues. The word 'plague' comes from the Latin *plaga*, meaning a stroke or a wound; the Oxford English Dictionary also defines it as an affliction, calamity or a general name for any malignant diseases 'with which men or beasts are stricken'. The word is, moreover, now becoming associated with some disorders that are not infectious, but which have increased in prevalence so rapidly that they have been likened to the historic plagues.

One such condition in humans is dementia, which has been called the 'twenty-first century plague', and indeed it is perhaps as greatly feared today as the classic plagues were in the past. At the first ever G8 Dementia Summit held in London in December 2013, the British Prime Minister, David Cameron, spoke about the global challenge of this condition, and described it in the following words:

> *It doesn't matter whether you're in London or Los Angeles, in rural India or urban Japan – this disease steals lives; it wrecks families; it breaks hearts and that is why all of us here are so utterly determined to beat it.*[1]

The most common form of adult dementia is Alzheimer's disease, named after Alois Alzheimer (1864–1915), who first described its symptoms and pathological character in a lecture in 1906. Alzheimer's disease is now one of the most distressing and debilitating conditions of the modern age. It is estimated that nearly one million Britons, and some 40 million people worldwide, suffer from this progressive neurodegenerative disease. Its symptoms, which include memory loss and disorientation, are familiar to everyone in a country such as ours, and it is rare for a day to pass without an article on the topic of dementia appearing prominently in the news. Moreover, there are few of us who do not know of a family member or friend with experience of this disease – either as a patient or as a carer – and who has not been deeply affected by its relentless nature and the despair that it engenders. And while its human costs are almost overwhelming, its financial burden – largely the costs of care and lost working days – is huge, estimated to be approaching £30 billion per annum in the United Kingdom alone.

In this chapter we give a broad chronological overview of the plagues of the past to try to understand how they have arisen, how they have affected societies over time, and how humanity has responded to the challenges of each new wave of deadly disease. We conclude that while each of these plagues is the result of diverse factors and pathogens, what is most striking is that they all reflect rapid changes in human activities and lifestyles that have occurred over the ages. In modern times, such

FIGURE 2.1 **Boils and Pustules**
This miniature, from the Swiss manuscript of the Toggenburg Bible, *c.* 1411, has traditionally been thought to depict a couple suffering from the 'buboes' of the deadly Black Death that swept through Europe in the mid-fourteenth century. Recently, scholars have questioned this, noting that this image of the Book of Exodus' sixth plague is not an accurate clinical depiction of bubonic plague and is more likely to represent a smallpox-like disease. [Source: Corbis/Betemann: 42-24063922]

changes are more rapid than ever. Indeed, new outbreaks of infectious diseases are ever threatening, but advances in scientific and medical knowledge have so far managed to avert catastrophes on the scale of those of the past. But it is the very success of modern science and medicine – in enabling large proportions of the population of the world to live to unprecedented ages – that has generated a huge increase in the incidence of afflictions such as dementia. We touch on some of the historical examples of major outbreaks of disease before turning to discuss how our increasing understanding of the origins of Alzheimer's disease has the potential to enable us to defeat this latest challenge to the health and welfare of the human race.

Plagues of the Ancient World

Many people will be familiar with the Biblical plagues of Egypt, about which there has been much speculation and many emotive artistic representations (See Figure 2.1). [2]

FIGURE 2.2 **The Plague of the Philistines at Ashdod**
This oil painting by the Flemish artist, Pieter van Halen (1612–87) of the 'Plague of the Philistines at Ashdod' is one of several graphic artistic representations of the plagues of antiquity painted during the Renaissance.
[Source: Wellcome Images: L0011603]

While there is little direct evidence for the ten afflictions that are described in the Book of Exodus, the explanation offered in the written account is that they were sent by God, although there have been more recent attempts to rationalise such afflictions as 'rivers turning to blood' in scientific terms. Another Biblical plague was the Plague of the Philistines (See Figure 2.2), where again the explanation given was the wrath of God:

> *... after they had carried [the ark] about, the hand of the Lord was against the city with a very great destruction: and he smote the men of the city, both small and great, and they had emerods in their secret parts.*[3]

As 'emerods' are better known as 'haemorrhoids' one can understand the distress that this particular outbreak of sickness must have caused. Other catastrophes of the distant past include the Plague of Athens (*c.* 430–426 BC), the Antonine Plague (*c.* AD 165–180) and the Plague of Justinian (AD 541–543) that struck civilisations of the Greek and Roman world, carrying with them a huge toll of death and destruction and even changing the course of history. Procopius of Caesarea (*c.* AD 500–565) vividly described the horrors of the Plague of Justinian 'by which the whole human race came near to being annihilated' and highlighted its puzzling nature as it devastated Constantinople (now Istanbul) and other parts of the Byzantine Empire:

> *For much as men differ with regard to places in which they live, or in the law of their daily life, or in natural bent, or in active pursuits, or in whatever else man differs from man, in the case of this disease alone the difference availed naught.*[4]

Why, how and where these plagues emerged and which specific diseases were involved still remain a mystery but, assuming that they were infectious diseases, their spread can be attributed primarily to the effects of overcrowding in cities, particularly in times of war, as human populations began to increase and people migrated and took refuge from the enemy.

While scholars continue to debate the impact and cause of these plagues, recent DNA evidence makes it likely that the Plague of Justinian was the first recorded outbreak of 'bubonic plague' – whose name reflects the characteristic 'buboes' (from the Greek, *boubón*, 'groin' or 'swelling in the groin') that appeared on the bodies of its victims. The plague bacillus, *Yersinia pestis*, enters at the site of the bite of an infected flea and travels through the lymphatic system to the nearest lymph node where it replicates itself. The lymph node then becomes inflamed, tense and painful, and is called a 'bubo'. Besides bubonic plague, there are two other forms of the disease – pneumonic plague – a highly virulent form (when *Y. pestis* travels to the lungs) and septicaemic plague – the most deadly of all (when the infection spreads directly through the bloodstream without forming a 'bubo').

Plagues of the Medieval World

The 'Black Death' that struck Europe in the mid-fourteenth century (*c.* 1346–53), has etched itself into the memory of the Western world as the classic plague. Possibly triggered by environmental or climatic disturbances which enabled the disease to jump from its natural reservoir in wild rodents, such as gerbils or marmots, to black rats (the species known as *Rattus rattus*) which live in close proximity with humans, the Black Death spread from Asia westwards along increasingly widely used maritime and land trade routes, and again especially affected overcrowded and insanitary ports and inland settlements. The Black Death resulted in the deaths of between 25 and 50 million people in Europe, alone. It was the greatest demographic crisis of the medieval period and, in terms of the proportion of the population that was killed, the single most calamitous epidemiological event in all of human history.

The term 'Black Death' was coined only much later, the word 'black' referring possibly to the sheer horror of the pestilence (from the Latin, *atra mors*, which can mean 'terrible' or 'dreadful' death, the connotation of which was the 'black death') or, as some have suggested, to the blackened bodies of its victims. Contemporaries called the epidemic the 'Great Pestilence', the 'Great Mortality' or the 'Big Sickness'. They described a range of symptoms, including buboes – the size of eggs or even apples – on the groin and under the armpits, as well as blotches, boils, bruises, black pustules and the coughing up of blood, vomit and sputum.

The poignant accounts writers left behind ring with the terrible sorrows it brought in its wake. The Italian poet Petrarch (1304–74) expressed the perplexity and loneliness that must have haunted those who survived:

> *Where are our dear friends now? Where are the beloved faces? Where are the affectionate words, the relaxed and enjoyable conversations? What lightning bolt devoured them? What earthquake toppled them? What tempest drowned them? What abyss swallowed them? There was a crowd of us, now we are almost alone.*[5]

Giovanni Boccaccio (*c.* 1313–75), author of *The Decameron* (*c.* 1348–53) – set during the pestilence in Florence – also described its tragic consequences:

> *How many valiant men, how many fair ladies . . . breakfasted in the morning with their kinsfolk . . . and that same night supped with their ancestors in the other world!*[6]

Perhaps no other single 'plague' has so changed the world. The sheer scale of the Black Death in terms of the numbers killed over the course of just seven or so years resulted in huge demographic, social and economic upheavals. The impact on individuals, families and communities at the time hardly bears thinking about but survivors did adapt to new circumstances and in some countries the position in society of, for example, formerly landless labourers and rural tenants gradually improved for the better after the Black Death (addressed in Chapter 3 and Chapter 6). Indeed, historian John Hatcher reminds us of the complexities for England of the outcome of this calamitous event:

> *But it should always be remembered that the rising living standards and improved status that the ordinary folk came to enjoy were brought at the huge cost of a terribly high and unpredictable mortality . . . Such momentous mortality naturally had the potential to create confusion and disorder, but equally striking is the speed and power with which forces within society and economy moved to restore stability . . . Whereas the historian is struck by the continuities, contemporaries would have been overwhelmed by the scale of changes.*[7]

Plagues of the Early Modern World

Bubonic plague continued to strike over the following centuries, and wreaked havoc despite the imposition of measures that attempted to restrict the spread of disease, including cleaning up the filth of cities, as well as shutting up the infected in their homes or in pesthouses, and the introduction of quarantine – a word that comes from the Italian for 'forty days' (*quaranta giorni*) – the time that ships had to wait in port before delivering their potentially infectious passengers and cargoes.

The most famous (or infamous) outbreak of early modern plagues in Britain, some three centuries after the Black Death, was the Great Plague of London in 1665–6. This tragic and memorable event is thought to have killed 70,000–100,000 people, or one-fifth to one-quarter of the overcrowded population of insanitary London, causing panic and terror. 'Searchers' – frequently illiterate 'elderly matrons' – were sent to seek out the dead and ascertain the cause of death. Bills of Mortality recorded the numbers who had died of the plague and other diseases. Dead-carts travelled the streets at night with their drivers calling 'bring out your dead'; they carried away the corpses, filling up mass graves to the brim. Perfumes, along with prayers, wafted through the churches, which were described by Daniel Defoe (*c.*1660–1731) as 'like a smelling-bottle; in one corner it was all perfumes; in another aromatics, balsamics, and variety of drugs and herbs.'[8] Following traditional advice during times of epidemics of 'flee early, flee far, return late', many who could afford to do so left the City, including not only King Charles II and his Court, but also priests and physicians. As Samuel Pepys (1633–1703) wrote on 16 October 1665 in his famous *Diary*:

> *But Lord, how empty the streets are, and melancholy, so many poor sick people in the streets, full of sores, and so many sad stories overheard as I walk, everyone talking of this dead, and that man sick, and so many in this place, and so many in that.*[9]

Contemporary ideas to explain the cause of plague included supernatural phenomena, divine vengeance, contagious particles, miasmas and foul smells. In some towns 'plague doctors' were recruited to treat the sick – their protective costumes reflecting the idea that the disease was thought to be highly contagious (See Figure 2.3).

Remedies ranged from prayer and penance to quack medicines sold to gullible and desperate people, as well as persecution of those groups in society suspected to have brought about these events. Although by the eighteenth century widespread epidemics of plague no longer struck Western Europe (for reasons which are still debated, though isolation and quarantine measures may have helped), the disease continued to be a substantial problem in other parts of the world, especially Asia, and between the 1890s and 1940s resulted in 15 million deaths worldwide. For the first time the plague spread to North America, sub-Saharan Africa and Australia via increasingly rapid forms of ocean transportation.

The 'plague bacillus' (*Y. pestis*) was, however, eventually found to be the cause of the disease by Alexander Yersin (1863–1943), working in Hong Kong at the end of the nineteenth century during this so-called third pandemic of bubonic plague. Once the rat-flea-human cycle of plague was elucidated and confirmed in the early twentieth century, measures were taken to trap and kill the rats that harboured the fleas that carried and transmitted the infection to humans. By the second half of the twentieth

FIGURE 2.3 **A plague doctor**

A plague doctor dressed in a seventeenth century preventive plague costume. While many professionally trained physicians did their best to avoid the infection, some brave souls (including those lured by the lucrative pay offered to volunteers, given the risk of death involved) were specifically hired as 'plague doctors'. This iconic image of the plague doctor shows him fully garbed in a head-to-toe protective costume, with the long, beak-like nosepiece stuffed with aromatic substances to combat the stench associated with plague. A popular seventeenth-century poem describes the plague doctor's costume:

In Rome the doctors do appear,
When to their patients they are called,
In places by the plague appalled,
Their hats and cloaks, of fashion new,
Are made of oilcloth, dark of hue,
Their caps with glasses are designed,
Their bills with antidotes all lined,
That foulsome air may do no harm,
Nor cause the doctor man alarm,
The staff in hand must serve to show
Their noble trade where'er they go.

[Source: Wellcome Images: V0010642]

century, antibiotics became available to treat the plague, and there are now only around 1000 to 2000 reported human cases a year. *Y. pestis* has recently been identified in skeletal remains of medieval plague victims, confirming that this bacterium was the cause of the Black Death (as well as the Plague of Justinian).

In fact, bubonic plague is a classic example of a 'zoonosis' when a disease jumps from animals (in this case rodents) to humans, a concept also known as 'spillover'. As we shall see later, this mode of the emergence of diseases in the human population is now thought to be extremely common, with 'bird flu', 'swine flu', Ebola, HIV/AIDS, SARS and MERS-CoV as well recognised examples. Once seeded in the human population, however, there is also the possibility that such diseases can be spread directly by human-to-human transmission. In the case of plague, patients may develop the 'pneumonic' form of the disease when the bacilli enter the lungs. Pneumonic plague is highly transmissible via infective airborne droplets. Untreated pneumonic plague has a very high case-fatality ratio and, while thought to be the least common form of plague, some scholars have suggested that an airborne mode of transmission may have accounted for the devastating death toll of the Black Death. There are, however, still many puzzles and lively ongoing debates about the origins and epidemiological histories of the plague pandemics of the past – not least, as some media headlines have recently put it: Was the much-maligned black rat (*Rattus rattus*) really 'to blame' for the Black Death and repeated outbreaks of plague across Europe? [10]

Although bubonic plague was ravishing parts of Europe until the beginning of the eighteenth century, it did not reach the Americas until the early twentieth century when it struck a number of cities, including San Francisco, in short outbreaks. Other types of 'plagues', however, spread across the world in the early modern period as a consequence of a new era in human history during which travels between distant lands became increasingly common. The disease with the most widespread global effects at this time was smallpox, an airborne viral infection that probably dated back to the rise of the first agricultural settlements some ten thousand years ago. Like other so-called crowd diseases, smallpox depended on a sufficient density of population to spread from human to human and appears to have become increasingly virulent during the early modern period. In Europe it became known as the dreaded 'speckled monster', attacking princes and paupers alike. By the sixteenth century it accounted for up to 15 per cent of all deaths and it left those who survived pockmarked and scarred. As the historian Lord Macaulay (1800–59) commented:

> *... the small pox was always present, filling the churchyards with corpses, tormenting with constant fears all whom it had not yet stricken, leaving on those whose lives it spared the hideous traces of its power, turning the babe into a changeling at which the mother shuddered, and making the eyes and cheeks of the betrothed maiden the objects of horror to the lover.*[11]

Smallpox first crossed the Atlantic with the Spanish conquistadors when they arrived in Central and South America in the 1490s. There it spread rapidly as the indigenous American populations had never experienced this infection in previous times and had acquired no immunity against it. Other infectious diseases too, such as measles, were transported across the ocean. Historians have suggested that the effects of these so-called virgin soil epidemics contributed to the collapse of the Aztec empire in Mexico and the Inca civilisation in Peru, though unravelling the many complex reasons for the demographic decline of the indigenous populations still remains an area of historical debate.[12]

The transport and transfer of disease was, probably, not just a one-way transmission from Europeans to Native Americans. The so-called Columbian disease exchange is thought to have introduced syphilis (the 'Great Pox') to the Old World from the New World – though, again, this is a topic that is still much debated by medical historians. Nonetheless, we know that serious epidemics of this sexually transmitted bacterial disease began to occur in Europe from the late fifteenth century onwards.[13] As with other apparently 'new' diseases, its dramatic effects were alarming, in this case often leading to gross disfigurement and debilitating neurological symptoms before a painful death. One chronicler, Joseph Grünpeck (*c.*1473–*c.*1532) thus described syphilis:

> *In recent times I have seen scourges, horrible sicknesses and many infirmities affect mankind from all corners of the earth. Amongst them has crept in ... a disease which is so cruel, so distressing, so appalling that until now nothing so horrifying, nothing more terrible or disgusting, has ever been known on this earth.*[14]

With the opening up of the African continent and the horrendous transportation of slaves across the Atlantic, further exchanges in the disease pools of the Old and New Worlds took place. As explorers reached the Pacific, smallpox arrived in Australia in 1789, only a year after the first English settlement was established. It resulted in a similar demographic and psychological shock to the indigenous population as that in the Americas. It has been estimated that half of the aboriginal people who had contact with the newly arrived settlers died of this disease.

Ways of controlling diseases such as smallpox were, however, increasingly being explored. In the early eighteenth century, Lady Mary Wortley Montagu (1689–1762), the wife of the British ambassador to the Ottoman court, learned about the practice of inoculation during her time in Constantinople from the local population and introduced smallpox inoculation into Britain. This traditional method (which had been used in a number of countries from the tenth century AD) involved the introduction of pus from smallpox eruptions into healthy people, for example under the skin or into the nose, to induce a mild case of the disease and thence immunity. An even more exceptional method of preventing smallpox came with the discovery by Edward Jenner (1749–1823) of vaccination (using pus from cowpox – *vacca* is the Latin word for cow) at the end of the eighteenth century (See Figure 2.4).

Vaccination proved to be much safer and more effective than inoculation, and this practice was adopted around the world with extraordinary speed, showing that medical interventions could travel almost as rapidly as the diseases themselves.

Although there were objectors to vaccination – on ethical and religious grounds – and 'anti-vaccinationists' fought hard to suppress the practice (discussed in detail in Chapter 3), it offered the pox-ridden world a huge opportunity to defeat this ancient and appalling viral disease. Indeed, in 1806 President Thomas Jefferson (1743–1826) wrote to Jenner from America:

> *Medicine has never before produced any single improvement of such utility . . . You have erased from the calendar of human afflictions one of its greatest ... Future nations will know by history only that the loathsome small-pox has existed and by you has been extirpated.* [15]

FIGURE 2.4 **Smallpox vaccination**
In this oil painting of 1884 by Eugène-Ernest Hillemacher, Edward Jenner (1749–1823) is seen vaccinating a small boy. Jenner's introduction of vaccination to prevent smallpox in the late eighteenth century was a remarkable example of empirical ingenuity that was highly effective before the causative agent of the disease was known. Vaccination, based on the principles established by Jenner, led to the official announcement of the global eradication of this ancient scourge in 1980. Today, vaccination plays a major role in preventing many infectious diseases.
[Source: Wellcome Images: L0029094]

It did, however, take many years for smallpox to be 'extirpated'. But over a century and a half after Jenner's pioneering discovery, the increasing efficacy of smallpox vaccines, and their widespread availability, enabled the World Health Organization to mount a campaign to eradicate the disease (even without a cure). By 1979 the world was declared free of smallpox and in 1980 it was removed from the list of world diseases. Smallpox was the first disease ever eliminated by human ingenuity – an outstanding triumph in the history of medicine.

Plagues of the Victorian World

During the Industrial Revolution of the nineteenth century another major change in human behaviour took place as a result of mass migration to the rapidly expanding cities, where employment and production were concentrated in factories – the 'dark satanic mills' of Victorian Britain and, increasingly, of other industrialising nations. A series of 'pandemics' (from the Greek, *pan-*, 'all', and *demos*, 'people') swept through densely populated and polluted cities. The most dramatic of these diseases was Asiatic cholera that spread across the world from its heartland in the Ganges Delta in the Indian subcontinent. With expanding empires and towns and cities interlinked through long-distance trade networks the threat of a 'new' disease reaching the shores of Europe was terrifying, as one doctor emphasised when cholera first struck Britain in 1831/2:

> *Our other plagues were home-bred, and part of ourselves, as it were ... But the cholera was something outlandish, unknown, monstrous ... a terror which ... seemed to recall the memory of the great epidemics of the middle ages.*[16]

Cholera, with its dramatic symptoms of vomiting and rice-water diarrhoea and rapid death through dehydration, hit the insanitary cities of Europe. It added to the mortality toll of diseases such as typhus (a louse-borne bacterial disease which had devastated Napoleon's Army during his Russian campaign in 1812), and typhoid (a waterborne bacterial disease which is said to have killed Prince Albert, Queen Victoria's beloved husband, in 1861). Tuberculosis (an airborne bacterial disease, often known as the 'white plague' from the pallor of its victims) was also responsible for countless deaths at this time.

People realised that the appalling conditions and lack of sanitation were serious problems within cities such as London, as this heartfelt appeal printed in *The Times* in 1849 shows all too clearly:

> *Sur ... We live in muck and filth. We aint got no priviz, no dust bins, no drains, no water-splies, and no drain or suer in the hole place. The Suer Company ... take no notice watsomdever of our complaints. The Stenche of a Gulley-hole is disgustin ... if the Cholera comes Lord help us.* [17]

There was, however, no clue as to the real cause of cholera and all sorts of possible explanations were aired (See Figure 2.5).

Contemporaries debated whether the disease was caused by 'miasmas' from the foul air (in which case sanitary reforms were needed), or by 'contagious' particles, spread from person to person (which should be controlled by maritime quarantine measures). An editorial in *The Lancet* in 1853 questioned:

FIGURE 2.5 **Hunting for cases of cholera**
A London Board of Health searching for the cause of cholera when it first hit Britain in 1832. Edwin Chadwick (1800–90), the English social reformer, believed that *'all smell is disease'*. In 1854, John Snow (1813–58) recognised that cholera was spread by contaminated water, which led, ultimately, to practical measures of separating sewage from drinking water. In a recent UK poll of doctors and the public, 'sanitation' topped the list of the greatest fifteen medical milestones since the mid-nineteenth century.
[Source: Wellcome Images: V0010896]

> *What is cholera? Is it a fungus, an insect, a miasm, an electrical disturbance, a deficiency of ozone, a morbid offscouring from the intestinal canal? We know nothing; we are at sea, in a whirlpool of conjecture.*[18]

But, as ever, there were some people able to rise to such challenges and investigate the causes of the various diseases and the means of their prevention. A classic example involved the brilliant detective work by John Snow (1813–58) in London in the mid-nineteenth century; Snow recognised from the pattern of infection that cholera was carried by contaminated water, and famously requested the removal of the handle of a pump in Broad Street in London's Soho district to stop the spread of an outbreak of cholera. Such epidemiological studies demonstrated the need for effective public health and hygiene measures, which were subsequently of major significance in the control of a host of infectious diseases.

The recognition of the ways that diseases such as cholera could be transmitted, together with advances in laboratory research, led to a transformation in the understanding of infectious illnesses. Of seminal importance was the discovery of the existence of 'germs' in the late nineteenth century, beginning with the pioneering work of the French scientist Louis Pasteur (1822–95) and including the identification by the German scientist Robert Koch (1843–1910) of the bacteria that cause tuberculosis in 1882 and cholera in 1883–4. Ultimately, these medical milestones were followed by the development of drugs (antibiotics) able to kill specific bacteria, especially the introduction of penicillin and its derivatives into clinical medicine from the mid-twentieth century. Sadly, the uncontrolled use of antibiotics has resulted in the emergence of resistant strains of bacteria in the modern world with potentially disastrous consequences for the future, as we discuss below.

Plagues of the Modern World

It was a virus, not a bacterium, however, that led to the most lethal pandemic that the human race had ever experienced in a short space of time: the 'Spanish flu' of 1918–19, which is estimated to have killed over 50 million people across the globe, more than the number of people who died during the First World War (See Figure 2.6).

With the trauma of the war this appalling pandemic initially attracted much less attention than it would otherwise have done, although its relevance to modern outbreaks of disease has now generated tremendous interest in its nature. As an editorial in *The Times* of London remarked in December 1918:

> *Never since the Black Death has such a plague swept over the face of the world; never, perhaps, has a plague been more stoically accepted.* [19]

Like the great plagues of the past, the origins of this disease, which has been described by historians as *'the greatest single demographic shock that the human species has ever received'*, were unknown. In fact, influenza is probably an ancient disease and, like smallpox, only became a highly

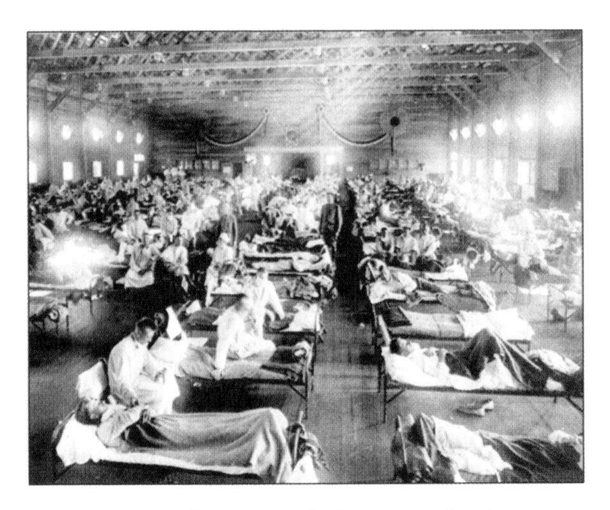

FIGURE 2.6 **The 1918-19 influenza pandemic**
The Spanish flu of 1918–19 was one of the most lethal pandemics in history, killing within a short time some 50–100 million people. This contemporary image shows soldiers from Fort Riley, Kansas, ill with Spanish influenza at a hospital ward at Camp Funston. [SOURCE: Wikimedia Commons (http://en.wikipedia.org/wiki/1918_flu_pandemic/media/File:CampFunstonKS-InfluenzaHospital.jpg]

transmissible disease once the density of human and domestic animal populations increased with the development of agricultural communities. The word 'influenza' was coined by the Italians – from the Latin, *influentia coeli*, or 'heavenly influence' – and introduced into the English language in the eighteenth century. The 1918–19 'Spanish flu' was so-called not because it started in Spain but because Spain was not a belligerent country and its press was not prevented by government censors from freely reporting its alarming impact when it struck there in May 1918. But why influenza afflicted so many parts of the globe, almost simultaneously, remains puzzling, as does the fact that it was so virulent and devastating and why, unlike seasonal flu (which mostly affects the elderly and very young), it primarily targeted young adults in the prime of life.

Efforts to prevent the spread of the disease – especially once its severity was evident and widely acknowledged – included prohibiting public gatherings, disinfecting streets and homes, sterilising water fountains, banning spitting and shaking hands (which, in some places, became a punishable offence), quarantining ships and enforcing the wearing of gauze masks. Some invoked folk remedies, such as carrying garlic, sulphur, cucumbers or potatoes to ward off infection, and, as in earlier times, any number of quack remedies and lung tonics were sold as 'sure cures'.

The discovery that influenza was caused by a virus was eventually made following the development of the electron microscope in the 1930s, and recently the advent of gene sequencing techniques has enabled the specific strain of the 1918–19 pandemic to be identified from preserved lung tissue, as well as from bodies of victims that had been buried in permafrost in Norway and Alaska. Vaccines and drugs have been developed to combat influenza, but we now know that the influenza virus can mutate and recombine genes, often in animal hosts, with frightening rapidity, resulting in outbreaks of forms of this disease for which existing vaccines are ineffective. Recent examples include certain types of avian and swine flu, the former ('bird flu') related to the strain giving rise to Spanish flu, adding to fears of the emergence of novel and highly contagious forms of this disease that are beyond our ability to control.

And again changes in human lifestyles make these fears all too real. Not only has the global population increased hugely in recent years along with the rise of 'mega-cities', but so, too, has the speed of travel and hence the transmission of contagious diseases. The world can now be circumnavigated in a couple of days or less, whilst a century or two ago it took many months. Indeed, it took several years for bubonic plague to spread across Asia to Europe in the fourteenth century, but the SARS (severe acute respiratory syndrome) virus travelled from Hong Kong to Canada in just a few days in 2003. SARS is a flu-like respiratory disease arising from a 'new' coronavirus and is yet another example of a zoonotic disease that crossed the species-barrier from animal hosts (bats and civets) to humans in crowded markets in China (See Figure 2.7).

As Ron Barrett and George Armelagos remind us:

> *The major themes governing our susceptibility to infectious diseases today are essentially the same as those of our ancient past: they are merely intensified by our massive populations, our cities, and technologies now at our disposal . . . modern humans are essentially "stone agers living in the fast lane."*[20]

FIGURE 2.7 **International health alerts during the SARS pandemic**
As a deadly outbreak of a 'new' disease, known as SARS (severe acute respiratory syndrome), threatened the globe in 2003, a number of measures were rapidly put into effect, such as quarantining the infected and monitoring the health of travellers at international borders to prevent its spread. SARS affected twenty-nine countries, reminding us that any infection is only a plane flight away.
[Source: Corbis DWF15-692817]

Another terrible and tragic affliction of modern times is AIDS (acquired immune deficiency syndrome), which is caused by the virus known as HIV (human immunodeficiency virus). HIV is also of zoonotic origin, having evolved from simian immunodeficiency viruses (SIV) of several non-human primate species. HIV was probably acquired by humans through the hunting and consumption of 'bush meat' in Africa. Estimates of the timing of its emergence in humans vary from fifty to one hundred years before its appearance in the United States in the early 1980s, when HIV/AIDS shocked the world as a new type of human disease. In 1983 the HIV retrovirus (an RNA virus whose DNA copy becomes incorporated into the host genome and so persists indefinitely as long as infected cells live) was identified and characterised.

Multiple epidemics of HIV/AIDS, with very different epidemiological and demographic patterns, have spread across the world. The impact of this new 'silent' plague, that typically takes several years to manifest itself as the human immune system is slowly destroyed, has been most devastating in the poorer countries of the world, especially sub-Saharan Africa. Here, this disease has often been spread by people who do not know they carry HIV, typically killing young adults and leaving behind millions of orphans. As one epidemiologist in Kampala, Uganda, wrote:

> *It all started as a rumour. Then we found we were dealing with a disease. Then we realized that it was an epidemic. And, now we have accepted it as a tragedy.* [21]

HIV is transmitted through sexual interactions, contaminated needles or infected blood products, and can also be passed from HIV-infected mothers to infants. The recognition of its public health dangers in the early 1980s led to a clamour for action, and campaigns for prevention and behavioural changes became the key targets of national and global health organisations. Around the world people were advised about the risks of contracting HIV/AIDS and the ways of avoiding or spreading it. The 'Stop AIDS' campaign became one of the biggest health-education drives the world had ever seen. Within just a few years a therapeutic breakthrough was made and AZT, the first drug licensed for this disease, was in use by 1987, having been fast-tracked through the US Food and Drug Administration (FDA). Soon the concept of a cocktail of antiretroviral drugs (combination therapy) was developed to combat the growing problem of anti-viral resistance, and now is so effective at containing the escape and replication of HIV that its use has become known as the 'Lazarus effect'.

Antiretroviral drugs were some of the first effective therapeutic compounds for any viral disease and their development represents a triumph of a massive and concerted programme of research involving academic laboratories and pharmaceutical companies – supported by both public and private funding as a result of intense lobbying in the United States in particular. Although not a cure, for those with access to these sophisticated drugs, HIV/AIDS is now seen as a chronic and manageable condition rather than a death sentence. Nevertheless, the disease has already claimed some 40 million lives and large numbers of people (probably some 35 million) are currently living with HIV infection. If left untreated, the slow destruction of the immune system leaves the body open to all sorts of life-threatening opportunistic infections and can, also, substantially worsen the manifestations of other co-infecting pathogens – especially tuberculosis. With increased availability of, and access to, antiretroviral drugs, particularly amongst those who urgently need them in the developing countries, there are hopes for an 'AIDS-free world'. Challenges still remain and scientists are continuing to search for the ultimate cure or a vaccine to combat this complex retrovirus that has become a fearful scourge of modern times (see also Chapter 5 by Stephen O'Brien).

Plagues of the Future World

Despite the tremendous progress that has been made in the fight against infectious diseases, there are still many 'ancient' diseases, particularly in the tropical parts of the world, which continue to plague humanity. These include malaria, tuberculosis, pneumonia and diarrhoeal diseases which are primarily associated with poverty, malnutrition, poor sanitation or a lack of access to a good standard of health care. Indeed, a number of viral, parasitic and bacterial infections, such as dengue fever ('breakbone fever'), onchocerciasis ('river blindness'), schistosomiasis ('snail fever'), African trypanosomiasis ('sleeping sickness') and Hansen's disease ('leprosy'), are now classed as 'Neglected Tropical Diseases' (NTDs) for which further attention is urgently needed. Moreover, the threat of emerging strains of infectious pathogens that are resistant to antimicrobial drugs is a problem that must be counteracted if the plagues of the past are not to re-emerge in even the most affluent societies.

The recent and severe outbreak in West Africa of Ebola (also known as Ebola haemorrhagic fever [EHF] or Ebola virus disease [EVD]) has made headline news. Like a number of other diseases we have discussed, Ebola is a zoonotic disease, thought to have originated in Africa from human contact with infected wildlife carriers (including bats). In August 2014, Ebola was labelled by the World Health Organization as an 'international health emergency'. This viral disease, once an initial 'spill-over' has occurred, can be spread directly from person-to-person by bodily fluids, and is an example of a relatively 'new' disease as it was first identified in humans only in 1976. It is so extraordinarily deadly that, during the height of the crisis in 2014, experimental treatments were administered that had not been subject to the usual degree of clinical scrutiny. In 2016, the emergence and spread of Zika virus, transmitted by mosquitoes, has also led to alarm bells ringing across the world. The threat of pandemics of infectious diseases with infected individuals rapidly criss-crossing around the globe has become the subject of vivid and shocking Hollywood productions with apocalyptic scenarios, such as *Outbreak* (1995) and *Contagion* (2011).

Notwithstanding such global attention and the local fear and panic they generate, these recurring themes remind us of the relationships of recent outbreaks of disease to earlier historical pandemics. Nevertheless, there have been incredible developments in humanity's response to outbreaks of 'plagues'. Over the past century or so, alongside public health, medical and surgical interventions and a phenomenal increase in our understanding of the biological, genetic and social determinants of diseases, there have also been major improvements in standards of nutrition, hygiene and health care in many (though, sadly, not all) parts of the world. A combination of such diverse efforts has enabled us to control many of the plagues of the past and, while we may still fear the threat of both 'old' or 'new' infectious diseases in our interconnected world, there are improvements being made in global surveillance systems that can track and monitor the spread of infections and, hopefully, prepare and protect us in the future from such global horrors of the past as the Black Death and the Spanish flu.

One of the most dramatic results of such medical and scientific advances is increasing life expectancy. For example, in the mid-nineteenth century the average life expectancy in the Western world was only about forty years, although, because of the shockingly high rates of child mortality, for those who survived until the age of twenty the average age of death was almost sixty years. Today, however, average life expectancy from birth in many parts of the world is eighty years or more. This extension of lifespan – in a large part as a result of combating infectious diseases – has, nevertheless, come at a price. Chronic diseases such as cancer and heart disease have become familiar and distressing afflictions of the modern age, but such conditions have fortunately received a very great deal of attention and funding. In 1969, US cancer specialist Sidney Farber (1903–73) is quoted as saying:

> *We are so close to a cure for cancer. We lack only the will and the kind of money and comprehensive planning that went into putting a man on the moon.*[22]

In 1971, the US Congress passed the National Cancer Act, which committed more funds and resources to cancer research and raised the profile of cancer enormously. A huge global industry, funded by governments, charities and pharmaceutical companies, grew up around what had become known as the 'War against Cancer'. Cardiovascular diseases (heart disease and stroke) have also attracted worldwide attention as serious and life-threatening conditions. And advances in heart surgery including transplantation, along with improvements in lifestyle risk factors, have had a major role in reducing early mortality from cardiovascular diseases.

There is a long way to go before the global toll of cardiovascular diseases and cancers (the two leading causes of death worldwide) is likely to be very significantly reduced, but we are clearly making substantial progress. By contrast, dementias, notably Alzheimer's disease, have silently crept up on us, and because of their very marked age dependence, we are only now beginning to acknowledge the serious burden that this 'twenty-first century plague' could have on us in the future (See Figure 2.8).

Although a small proportion of cases of Alzheimer's disease are 'early-onset' forms associated with specific genetic mutations, the large majority of cases are termed 'sporadic'. The risk of onset in these circumstances increases dramatically above the age of sixty-five, and it appears that over one third of people who have reached the age of eighty-five have at least some symptoms of this disorder. With the changing demographic patterns of modern societies – it has, for example, been estimated that more than half of all the people in human history who have reached the age of sixty-five are alive at the present time – the number of people suffering from this disease is likely to triple in the next forty years unless effective treatments are discovered. Without such treatments, however, the disruption to family and social life caused by Alzheimer's disease will become increasingly serious, as will the financial consequences; the cost

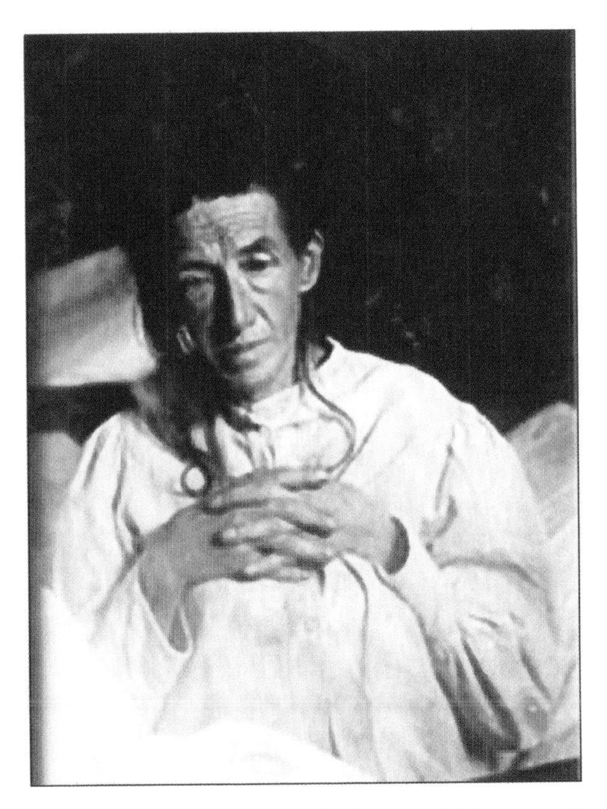

FIGURE 2.8 **The first recorded case of Alzheimer's disease**
In 1901 Auguste Deter was admitted to the Institution for the Mentally Ill and for Epileptics in Frankfurt, Germany, suffering from memory loss, delusions and other psychological problems. Alois Alzheimer examined her and noted that she was suffering from a 'peculiar disease'; after her death in 1906 he conducted an autopsy where he saw dramatic shrinkage of the brain and abnormal deposits around nerve cells. Such deposits had previously been observed in brain tissue by Rudolf Virchow (1821–1902) and had been named 'amyloid' because they stained with dyes used to detect starch (*amylum* in Latin); subsequently, it was shown that these deposits are in fact composed of aggregated protein molecules. As Deter was only fifty-one years old when Alzheimer first examined her, it is likely that she was suffering from an early onset form of disease caused by a genetic mutation, a conclusion that has recently been confirmed by analysis of preserved brain tissue using modern technologies.
[Source – Google Images]

of the disease in the United States alone is predicted to pass $1 trillion per annum by 2050. And dementia is not just an affliction of the richer nations. The World Health Organization estimates that some 70 per cent of the nearly 140 million sufferers of Alzheimer's disease anticipated by 2050 will be in low- and middle-income countries, putting an incalculable burden on already stretched health care budgets. [23]

Like most emerging diseases, the cause of Alzheimer's disease was at first unknown. In words reminiscent of those given above in the context of cholera in the mid-nineteenth century, the organiser of a research workshop in 1984 made the following comment:

> *We are still uncertain whether Alzheimer's disease is a specific, discrete,* qualitative *disorder such as an infectious process, endogenous or exogenous toxic disorder, or biochemical deficiency, or whether it is a* quantitative *disorder, in which an exaggeration and acceleration of the normal aging processes occur and dementia appears when neural reserves are exhausted and compensatory mechanisms fail.* [24]

Since then, however, much has been learned and it is now clear that Alzheimer's is a specific disease and not an inevitable consequence of ageing. Moreover, it has emerged that Alzheimer's disease is one of a family of some fifty types of medical conditions that are collectively known as 'protein misfolding diseases', which have only come to prominence in recent years. These disorders include other well-known forms of neurodegenerative conditions such as Parkinson's and Huntington's diseases, which like Alzheimer's disease are age-related, and also quite different disorders, notably type II diabetes, a condition that is itself proliferating rapidly, with more than 300 million affected individuals worldwide at present. This chronic affliction has resulted because of a different type of lifestyle change – in this case, lack of exercise and alterations in diet leading to the prevalence of obesity. It is predicted that the huge rise in the number of cases of type II diabetes could soon bring to an end the present steady increase in life expectancy in many parts of the world.

It turns out that this family of diseases is the result of the effects of 'pathogens', but unlike the historic plagues these pathogens are not external agents such as bacteria or viruses, but are generated within our bodies because of the aberrant behaviour of some of the very molecules on which our lives depend. Living systems are highly complex molecular machines made up of trillions of cells of many different types, from erythrocytes (red blood cells) and lymphocytes (white blood cells) to hepatocytes (liver cells) and neurons (nerve cells). These cells carry out all the processes necessary for life (for example, enabling us to breathe, fight disease, digest food and process information). The agents that enable all these functions to occur are proteins, long chains of chemical building blocks called amino acids; by varying the number and order of these building blocks (defined by the inherited information contained within our DNA) our bodies are able to make tens of thousands of different types of protein molecules, which together possess all the functional properties that are needed for life.

In order for proteins to gain their functions after the building blocks are joined together, however, they need to fold up into highly specific, compact, and often intricate, shapes. The process of protein folding is a complex and fascinating one, and although carefully regulated in our cells it sometimes goes wrong and results in 'misfolded' proteins. These misfolded proteins not only fail to function properly, but can also stick to each other to form clumps of molecules (known as amyloid) that can be highly toxic to cells, thereby disrupting the well-managed processes that ensure good health. Each disease is associated with the conversion of a specific type of protein molecule into the amyloid state, and there are therefore inherent similarities between the molecular origins and means of progression of this whole class of diseases. Indeed, to ensure that misfolding is not normally a problem, we have a wide range of protective mechanisms within our bodies to detect and degrade misfolded proteins and prevent pathogenic amyloid structures from being formed. These protective systems work very well under the types of conditions that have existed throughout human evolution. But in the last century or so all this has changed; in affluent parts of the world we live on average very much longer than ever before and have more than enough to eat without needing to exert the degree of effort that was previously required to acquire and produce food.

Under these circumstances, our protective mechanisms have an increasing tendency to fail, enabling misfolded proteins to accumulate,

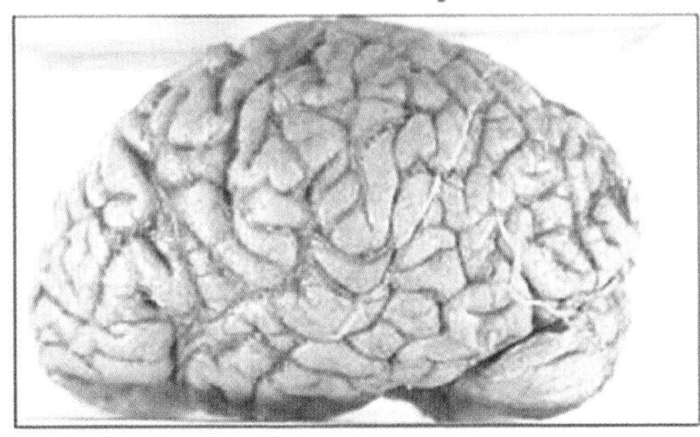

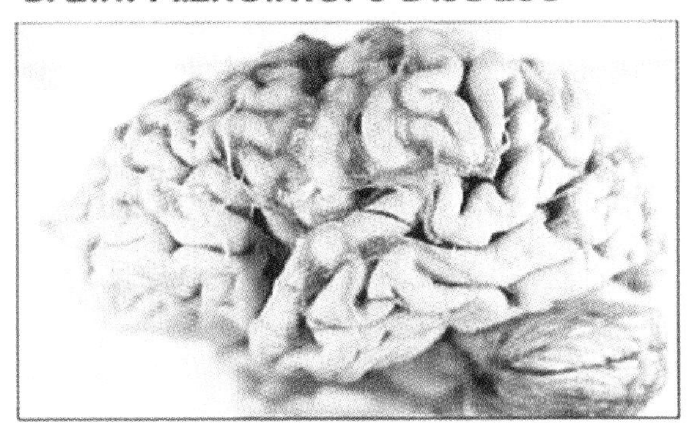

FIGURE 2.9 **Changes in the brain associated with Alzheimer's disease** Alzheimer's disease is characterised by loss of memory and cognitive impairment that results from the death of vital cells (neurons) in crucial areas of the brain. Such changes can be attributed to the misfolding and aggregation of proteins, ultimately forming amyloid plaques and also neurofibrillary tangles. The consequences of advanced neuronal loss and tissue atrophy are evident in the marked shrinkage of the brain shown in the lower image. [Source: The Sanders-Brown Center on Aging, University of Kentucky]

losing their normal functions and also generating pathogenic amyloid structures that can progressively damage or destroy the cells in their vicinity. If these pathogenic agents are in the brain, this phenomenon can give rise to dementia; indeed the clumps of misfolded proteins can accumulate to form the well-known 'amyloid plaques' in the brains of sufferers of Alzheimer's disease that are characteristic of this condition (See Figure 2.9).

A consequence of our extended life-spans, resulting in large part from the successful control and treatment of diseases that had ravaged human populations in the past, is therefore the proliferation of the number of cases of dementia, in many ways as terrible an affliction as many of the plagues of the past. In the words of Margaret Chan, Director General of the World Health Organization, in 2013:

> *I can think of no other condition that places such a heavy burden on society, families, communities, and economies. I can think of no other condition where innovation, especially breakthrough discoveries, is so badly needed.*[25]

Fortunately, our current understanding of the key elements of Alzheimer's disease, along with the continuing advancement of scientific and medical knowledge, is such that we have every opportunity to take control of our own future and ensure that we do not experience a catastrophe on the scale that afflicted so many of our ancestors.

Conclusions

We have discussed in this chapter how 'plagues' are very generally associated with rapid changes in human lifestyle and behaviour. Throughout history, humanity has responded by striving to discover the underlying causes of disease, and, armed with such understanding, find effective methods of prevention and even cures. In the case of infectious diseases, success has been demonstrated time and time again. The global eradication of smallpox has been the outstanding triumph and, indeed, following further vaccination campaigns, we now optimistically await the eradication of another serious viral disease – polio – from the globe. Yet as Albert Camus (1913–60) reminded us in his 1947 allegorical story, *La Peste* ('The Plague'), set in the Algerian port city of Oran (and discussed by Rowan Williams in Chapter 9), we must not be too complacent:

> *... the plague bacillus never dies or vanishes entirely ... and that perhaps the day will come when ... the plague will rouse its rats and send them to die in some well-contented city.*[26]

In addition to the emergence of old or new threats, as we respond with new medical advances, strains of bacteria, parasites and viruses emerge that are resistant to even our best antimicrobials or vaccines, and are a cause of huge international concern. It is essential that urgent action is taken to reduce disease risks by simple preventative measures, such as improving standards of hygiene, and to use existing antimicrobials and vaccines more prudently. Major efforts are underway to develop new classes of therapeutics to combat the threat of diseases that possess the potential to cause global pandemics. As Sally Davies, the UK Government's Chief Medical Officer, said recently:

> *If we don't take action, then we may all be back in an almost 19th Century environment where infections kill us as a result of routine operations. We won't be able to do a lot of our cancer treatments or organ transplants.* [27]

Nevertheless, we are now in a position where we can take advantage of the great advances that have taken place, and continue to take place, in scientific and medical knowledge. Attempts to develop drugs to combat Alzheimer's disease and related disorders have so far been unsuccessful because of the lack of knowledge of their molecular nature and inherent origins. Over the last twenty years or so our understanding of the underlying causes of these diseases, however, has been transformed through innovative research programmes that bring together scientists from a wide range of physical, biological and medical disciplines. The international scientific community is now in the process of devising rational diagnostic and therapeutic strategies for protein-misfolding disorders including Alzheimer's disease. We are optimistic that very significant breakthroughs will emerge from these efforts, just as the efforts of medical pioneers in earlier times resulted in ways of confronting the plagues of the past. Nevertheless, a great deal of highly innovative research will be needed to translate fundamental understanding of disease into effective drugs.[23]

Of crucial importance in securing such breakthroughs is a greater general understanding of the threat that these diseases, with their common underlying origins, albeit influenced by a complex mix of genetic, lifestyle and age-related risk factors, poses for our present populations. It is vital that there is a wider recognition that this threat will increase yet more dramatically in the future unless the global community provides substantial human and material resources specifically to prevent and combat these disorders. When a 'war' was declared on cancer in 1971, it resulted in huge increases in research funding by public and private organisations alike. This investment and dedication has had a very significant effect in developing new and powerful therapies. But funding for research into future plagues such as dementia is even now only a fraction – less than 10 per cent – of that spent on cancer. We have the knowledge and skills to combat dementia, but not yet the resources to do so quickly enough to avert the 'time-bomb' that is ticking ever more rapidly. As the late Terry Pratchett, author and former Patron of Alzheimer's Research United Kingdom, said in 2008:

> *We are facing . . . a worldwide tsunami of Alzheimer's and other dementia diseases. There's only two ways it can go: researchers, with as much help you can give them, may come up with something that reduces the effects of this dreadful, inhuman disease, or we will have to face the consequences of our failure to prevent the final years of many of us being a long bad dream. The strain on carers and their support is bad enough now; before very long the effects on the health service and society itself, will be unbearable.* [28]

History tells us that great civilisations have in the past suffered terribly from the effects of plagues, but that they also adapted and responded with remarkable tenacity. The increasing recognition of dementia as a rapidly developing 'plague' of the modern era is at last drawing attention to one of the greatest future global challenges. It is vital that we act at once, with real determination, to address this challenge, building on the progress that has been made in understanding the origins of the disease, and so grasp the opportunity of the moment for the sake of our children, and indeed for all future generations of the human race.

References

1. G8 Dementia Summit: Prime Minister's Speech, London, 11 December 2013; www.gov.uk/government/publications/g8-dementia-summit-aggreements/g8-dementia-summit-declaration. Accessed December 2013.
2. Jones L., Nevell R. (2016) Plagued by doubt and viral misinformation: The need for evidence-based use of historical disease images. *The Lancet Infectious Diseases* 16: 235–40.

3. King James Bible, 1 Samuel, chapter 5, verse 9.
4. *History of the Wars, Books I and II (of 8), by Procopius*, transl. H.B. Dewing, www.gutenberg.net. Accessed December 2013.
5. Letter from Francesco Petrarch, 1349, quoted in Rosemary Horrox, *The Black Death* (Manchester University Press, 1994) p. 248.
6. Boccaccio G. *The Decameron* (1348–53), transl. by John Payne, Day the First, www.gutenberg.net. Accessed December 2013.
7. Hatcher J. (2009) *The Black Death: The Intimate Story of a Village in Crisis, 1345–1350*. Phoenix, pp. 321, 224–5.
8. Defoe D. *A Journal of the Plague Year* (1722; reprinted 1970 Penguin) p. 218.
9. *The Diary of Samuel Pepys* volume VI – 1665 (edited by Robert Latham and William Matthews, Harper Collins, 1995), 16 October 1665.
10. www.bbc.co.uk/news/science & environment; 24 February 2015. Accessed February 2015.
11. Macaulay T.B. (1848)*The History of England from the Accession of James II*, volume IV, www.gutenberg.net. Accessed December 2013.
12. Catherine Cameron, Paul Kelton, Alan Swedlund, (eds.) (2015) *Beyond Germs: Native Depopulation in North America*. University of Arizona Press.
13. Arrizabalaga, J., Henderson J., French, R. (1997)*The Great Pox: The French Disease in Renaissance Europe* . Yale University Press.
14. Quoted in Porter R. (2003) *Blood and Guts: A Short History of Medicine*. Penguin Books, p. 13.
15. Jefferson T. Letter to Dr. Edward Jenner, 14 May 1806.
16. Quoted in 'Cholera and the Thames' website (Heritage Lottery Fund, 2015), www.choleraandthethames.co.uk/. Accessed December 2013.
17. *The Times*, 5 July 1849.
18. Wakley T. *The Lancet*, 22 October, 1853, p. 393.
19. *The Times*, 18 December 1918, p. 5.
20. Barrett R., Armelagos G. (2013) *An Unnatural History of Emerging Infections*. Oxford University Press, pp. 109, 111.
21. Quoted in Iliffe J. (2006) *The African AIDS Epidemic: A History*. James Currey Ltd, Oxford, p. 25.
22. From an advertisement in the *New York Times* and *Washington Post*, December 1969, declaring 'Mr. Nixon: You can cure cancer.' Accessed December 2013.
23. Dobson C.M. (2015) Alzheimer's disease: Addressing a twenty-first century plague. *Rendiconti Lincei* 26:251–62.
24. Quoted in Lock M. (2013) *The Alzheimer Conundrum: Entanglements of Dementia and Aging*. Princeton University Press, p. 54.

25. G8 Dementia Summit: Dr Margaret Chan, WHO Director-General, London, 11 December 2013; www.who.int/dg/speeches/2013/g8-dementia-summit/en/. Accessed December 2013.
26. Camus A. *The Plague* (first published in French, *La Peste*, 1947; Penguin Modern Classics edition, 2013), pp. 237–8.
27. www.bbc.co.uk/news/health; 11 March 2013. Accessed March 2013.
28. www.bbc.co.uk/news/health; 26 November 2008. Accessed December 2013.

Further Reading

Aberth J. (2011) *Plagues in World History*. Rowman and Littlefield Publishers.

Barrett R., Armelagos G. (2013) *An Unnatural History of Emerging Infections*. Oxford University Press.

Barry J. (2009) *The Great Influenza: The Story of the Deadliest Pandemic in History*. Penguin.

Bynum W., Bynum H. (eds) (2011) *Great Discoveries in Medicine*. Thames & Hudson.

Byrne J. (ed.) (2008) *Encyclopedia of Pestilence, Pandemics, and Plagues*. Greenwood Press.

Campbell B. (2016) *The Great Transition: Climate, Disease and Society in the Late Medieval World*. Cambridge University Press.

Chit F., Dobson C. (2006) Protein misfolding, functional amyloid, and human disease. *Ann. Rev. Biochem.* 75:333–66.

Cliff A., Smallman-Raynor M. (2013) *Oxford Textbook of Infectious Disease Control: A Geographical Analysis from Medieval Quarantine to Global Eradication*. Oxford University Press.

Crawford D. (2013) *Virus Hunt: The Search for the Origin of HIV*. Oxford University Press.

De Kruif P. (1926; republ., 1996) *Microbe Hunters*. Harcourt, Inc.

Dobson C. (1999) Protein misfolding, evolution and disease. *Trends Biochem. Sci.* 24:329–32.

(2002) Protein-misfolding diseases: Getting out of shape. *Nature* 418:729–30.

Dobson M. (1997) *Contours of Death and Disease in Early Modern England*. Cambridge University Press.

(2007) *Disease: The Extraordinary Stories Behind History's Deadliest Killers*. Quercus.

(2013) *The Story of Medicine: From Bloodletting to Biotechnology.* Quercus.
(2015) *Murderous Contagion: A Human History of Disease.* Quercus.
Doherty P. (2013) *Pandemics: What Everyone Needs to Know.* Oxford University Press.
Draaisma D. (2009) *Disturbances of the Mind.* Cambridge University Press.
Garrett L. (1994) *The Coming Plague – Newly Emerging Diseases in a World Out of Balance.* Penguin.
Gilbert M. (2014) *Yersinia pestis*: One pandemic, Two pandemics, Three pandemics, More?. *Lancet Inf. Dis.* 14:264–65.
Glynn I., Glynn J. (2004) *The Life and Death of Smallpox.* Cambridge University Press.
Green M. (ed.) (2015) *Pandemic Disease in the Medieval World: Rethinking the Black Death.* ARC Medieval Press.
Halliday S. (2011) *The Great Filth: Disease, Death and the Victorian City.* The History Press.
Hardy A. (1993) *The Epidemic Streets: Infectious Disease and the Rise of Preventive Medicine, 1856–1900.* Oxford University Press.
Harrison M. (2012) *Contagion: How Commerce Has Spread Disease.* Yale University Press.
Hatcher J. (2008) *The Black Death: The Intimate Story of a Village in Crisis, 1345–1350.* Phoenix.
Heeney J., Dalgleish A., Weiss R. (2006) Origins of HIV and the evolution of resistance to AIDS. *Science* 313:462–66.
Honigsbaum M. (2009) *Living with Enza: The Forgotten Story of Britain and the Great Flu Pandemic of 1918.* Macmillan.
Horrox R. (1994) *The Black Death.* Manchester University Press.
Hotez P. (2008) *Forgotten People, Forgotten Diseases: The Neglected Tropical Diseases and Their Impact on Global Health and Development.* ASM Press.
Iliffe J. (2006) *The African AIDS Epidemic: A History.* James Currey Ltd, Oxford.
Knowles T., Vendruscolo M., Dobson C. (2014) The amyloid state and its association with protein misfolding diseases. *Nature Rev. Mol. Cell Biol.* 15:384–96.
Mukherjee S. (2011) *The Emperor of All Maladies: A Biography of Cancer.* Fourth Estate.
Oldstone M. (2010) *Viruses, Plagues, and History: Past, Present and Future.* Oxford University Press.
Pepin J. (2011) *The Origins of AIDS.* Cambridge University Press.

Piot P. (2013) *No Time to Lose: A Life in Pursuit of Deadly Viruses.* W.W. Norton.

Porter R. (1997) *The Greatest Benefit to Mankind: A Medical History from Antiquity to the Present.* Harper Collins.

Quammen D. (2012) *Spillover: Animal Infections and the Next Human Pandemic.* The Bodley Head.

Rhodes J. (2013) *The End of Plagues: The Global Battle Against Infectious Diseases.* Palgrave Macmillan.

Schmid B. *et al.* (2015) Climate-driven introduction of the Black Death and successive plague reintroductions into Europe. *Proc. Nat. Acad. Sci. USA.* 112:3020–25.

Sherman I. (2007) *Twelve Diseases that Changed Our World.* American Society for Microbiology.

Slack P. (2012) *Plague: A Very Short Introduction.* Oxford University Press.

Stepan N. (2011) *Eradication: Ridding the World of Diseases Forever?* Reaktion Books.

Wagner D. *et al.* (2014) *Yersinia pestis* and the Plague of Justinian 541–543 AD: A genomic analysis. *Lancet Inf. Dis.* 14:319–26.

Wolfe N. (2011) *The Viral Storm: The Dawn of a New Pandemic Age.* Allen Lane.

Zimmer C. (2011) *A Planet of Viruses.* University of Chicago Press: Chicago, Illinois, USA. ISBN: 978-0-226-98335-6.

Zinsser H. (1935; republ., 2008) *Rats, Lice and History.* Transaction Publishers.

3 Plagues and Medicine

LESZEK BORYSIEWICZ

Throughout history many plagues have struck mankind. For nearly as long, mankind has endeavoured to fight the diseases underlying plagues. This chapter highlights the importance of integrating academic disciplines to sustain the considerable progress that we have made in the control of infectious diseases. It is also a personal reflection, which comes with a health warning. Out of necessity it is selective and not comprehensive. Scientific discoveries have changed the field of infectious diseases rapidly. New technologies constantly impact on the approaches that can be brought to bear on the problem. However, the study of the prevention of infectious diseases is not just about the biology underlying disease, but also about understanding society and individual behaviour. Ultimately, this is the conclusion I wish to deliver: the issues raised by our attempts to prevent plagues affect every member of society, not just the few fascinated by pathogens or our response to them.

Plagues' Impact on Society

Most appropriately, the start of my analysis is 1347 and The Plague, known as the Black Death, that hit Europe. The historical details of The Plague are discussed in the preceding chapter 'Plagues and History' by Mary and Christopher Dobson. However, the importance of The Plague in Europe was that it was the first well-documented large-scale population pandemic. It affected everybody and spread to all parts of Europe, but leaving certain geographical pockets less affected than others.

To examine how to prevent the spread and impact of pathogens, it is important to consider the impact of The Plague on medieval society. Pieter Bruegel in his masterpiece *The triumph of death* (c.1562) depicts every conceivable grisly way of ending your life in medieval times.

From scaffolds to being tortured on the wheel, a variety of ways of being decapitated or having other gruesome injuries inflicted on you are all there. But if this reflects the mind of medieval man, right in the centre is The Plague, with the characteristic plague cart filled with bodies.

So here is a perception of how plague affects individuals and in particular medieval man. Despite the fact that life was short and very hard by today's standards, the plague was perceived as a terrifying terminal event.

How did medieval man respond to this fearsome event? There were three possible options:

The first instinct, if you could afford it and the feudal system allowed it, was to run. Most of us faced with this scenario would choose the escape option. However, The Plague follows; in fact this movement possibly aided its transmission. Secondly the community could 'lock down'. Because the cause of the pestilence was unknown, and there was no concept of germ theory or other plausible explanation the community could lock itself inside a castle or city and pull up the drawbridge and isolate itself. This approach might work. Although the cause is unknown, isolation stops others who might be affected coming into contact with you. However, this approach has wider ramifications as a conscious decision has been made to restrict the rights of every individual to enter or leave. This may not have affected medieval man dramatically as drastic restrictions on mobility were an integral aspect of the feudal system. However, in today's times this is a major imposition in democratic societies. Nevertheless it remains an option to be applied at times of urgent need and has been successfully incorporated into the control of epidemics such as SARS [discussed in Chapter 5 by S.J. O'Brien]. Initially there were also signs it could be used with Ebola. However, as discussed in chapter one, the containment of Ebola in urban centres with insufficient health care infrastructure, poor education and communication has proven extremely difficult, especially when frightened exposed individuals flee potential quarantine in a highly mobile modern world. But whenever isolation or forced quarantine is used we must recognise this approach for what it is: society restricting the rights of an individual for the greater benefit of the population and therefore not a decision that can or should be taken lightly.

The third approach is less palatable but still evident even in today's more enlightened times. Often, without good reason, the response is 'blame someone, blame anyone'! In the mind-set of the middle ages, this disaster was a visitation from God to an ungodly sinful community. The church, dominant in influencing opinion, played up to it, and why not? Purification and atonement for the sins of man required chastisement, which was literally interpreted in the development of the flagellant movement; chastise and prostrate yourselves before God, surely he'll make this pestilence go away. He did not.

As an alternative, groups within communities may indulge in sectarian violence; blame somebody else, especially the outliers in the community. In many countries this happened to be the Jews; as they also held most of the debts of medieval society, it enabled you to literally 'kill two birds with one stone'. Not a savoury way for humanity to behave but, in the face of natural disaster, one that periodically recurs today. A current example in the context of the West African Ebola outbreak is that the deadly virus was actually a Western plot; a rumour often circulated in some communities. International Aid workers were often stoned and in some instances killed when they approached villages to help provide instruction and support.

What other measures could be implemented? These will largely depend on what people believe causes The Plague. During medieval times there was no concept of microorganisms, so other ideas come to the forefront. These are now known to be erroneous, but the response to prevention of spread was not illogical if the underlying premise of causation at the time was accepted. Medieval towns were smelly. Dead bodies smell so could 'miasmas' be important? In the context of plague the first to protect themselves (understandably) were the doctors. Personal protection became a high priority when visiting afflicted individuals. So the classical plague doctor's uniform came into being. Inside you would breathe in fresh, scented air through a herb-filled beak. Although the underlying premise of causation has changed, the importance of personal protection in contact with infectious cases is just as valid today. This is highlighted in the recent treatment of Ebola patients, where several medical personnel have contracted the disease despite advanced personal protection and strict protocols.

The corollary of the miasma hypothesis means that transmission can occur from the dead and disposal becomes a key issue. This resulted in the widespread usage of 'plague pits' in many countries with a remarkable uniformity of practice. Bodies collected in the characteristic plague carts that have almost become iconic symbols of the Black Death (Figure 3.1) were thrown into deep pits, with minimal religious observance, and lime usually added. Fascinatingly, it is from these pits and from the DNA extracted from teeth that we have been able to show that in fact the Black Death was due to *Yersinia pestis*, the pathogen that causes bubonic plague.

If the underlying premise is a 'religious curse' then protection with crosses and other religious symbols was common. These were used in marking doors of individual affected households. Locking yourselves away was often led by the church – famously as a self-sacrifice to protect

FIGURE 3.1 Pieter Bruegel *The triumph of death* c.1562. (Museo del Prado, Madrid, Spain).

others in Eyam, during a later plague. Some concept of person-to-person spread was evident, otherwise why isolate and quarantine during outbreaks. Such measures require coordination and enforcement and the development of the concept of 'public protection', where society imposes its will on actions of individuals. The orders of the 1665 Plague in London restricted the opening of theatres or of any public places in order to try to prevent spread of The Plague. While the impact on various strata of medieval and seventeenth-century society was variable, as a concept it laid the foundation for future public health interventions.

The premise of isolating sporadic cases to prevent wider spread of an epidemic led to the development of pesthouses, probably first introduced in Milan. Whether this was a totally novel concept or borrowed from isolation imposed from biblical times in relation to leprosy remains an open question. These pesthouses can still be identified on the edges of villages in Britain.

Rudimentary public health measures have been introduced whenever plagues occurred, even in the absence of a credible biomedical construct of the disease. But most large-scale plagues that have affected mankind leave their scars and subsequent consequences on society and the Black Death was no exception.

First, the huge demographic change had a direct impact. In Europe, mortality rates, in cities that measured it, varied dramatically. In Florence and Venice over 50 per cent of the population died but in Milan hardly anyone did. To this day we do not understand the basis for these variations. Host genetic variation differences in human genes that cause us to respond differently to infections is a distinct possibility, as well as pesthouses, which became very popular following Milan's introduction of this intervention. Even in medieval times there was communication, sharing of best practice and adoption of different approaches learned from other centres.

But the sheer scale of mortality changed rural populations, and the need to replace a lost workforce ensured that there was subsequent greater mobility of labour than previously possible. This led to a burgeoning of city populations and the end of the feudal system (see Chapter 6, 'Plagues and Economic Collapse'). So plague on this grand scale changed the organisation of medieval society.

It is also possible that plagues contributed to a change of attitude in beliefs. The belief was that pestilence was associated with 'sin', and the Church promoted this view. However, acquiescence to the edicts of the Church made no difference, and this directly challenged the authority promoting that belief. Furthermore, the church hierarchy did not remain anywhere near to afflicted areas. The clergy often locked themselves away; monks and priests closed off the abbeys. However, friars remained within the population. What was interesting is the suggestion that friars became more respected as they stayed and took their chances with society. In time the reputation of the church recovered to an extent but Ziegler, and others, hold the view that its social position never achieved the same level of respect. They extrapolate that the Black Death was part of the underlying sociological change that ultimately led to the Reformation.

Plagues, even today, can significantly affect demographic profiles of countries. One example is the impact of HIV and AIDs in South Africa, where even in 2013 approximately 30 per cent of all deaths were attributed to HIV/AIDS. Infectious agents and plagues still have the capacity to change populations.

Society's response to the prevention of infectious diseases is not always rational. There are logical interventions that can be based on an erroneous premise concerning causality. If the nature of the causality is not evidence-based on a thorough understanding of biology or epidemiology, it could be flawed. The consequent failure of an intervention based on that premise is itself logical and can be readily excused. However, this logical approach does not always happen in practice and opposition to preventative measures often flies in the face of bio-medical evidence. The reaction and opposition to the vaccination against smallpox, for example, was an intervention that subsequently eradicated the disease, and warrants examination.

Example of Smallpox

Smallpox is caused by a DNA poxvirus, and was previously feared as much as the plague. Historically, it was a disease that was widely prevalent with periodic epidemics and variable mortality throughout the eighteenth century in the United Kingdom. Whether this variability was in reality due to Variola major or a minor circulating variant, Variola (alastrim) minor, which has a mortality of only around 1 per cent, we do not know for certain. What is known is that in 1974, even with modern supportive medicine, in India the mortality rate in patients infected with Variola major was approximately 26 per cent.[1]

Transmission requires close contact and is mostly due to face-to-face spread. The secondary attack rate i.e. the number of people who will be affected if they are unvaccinated from a single point case coming into a close household, is about 58 per cent. Spread by fomites (that is clothing and other material) did occur, but it was not the major means of transmission. However, there is an asymptomatic incubation period for part of which individuals are infectious.

Jenner, using cowpox, introduced vaccination and gradually this replaced an older practice of variolation, where small amounts of infected material from cases of smallpox were inoculated into healthy children and adults. But how was vaccination promoted as a clinical practice and how did the public react to this preventative measure? To exemplify initial public opinion it is worthwhile looking at the depiction of vaccination in cartoons.

These two cartoons separated by only four years, present vaccination as distinct opposites. The classical Gillray cartoon of 1808 (Figure 3.2) shows the consequence of vaccination as cows emerging from those who are vaccinated, implying a negative view of the risks associated with this novel intervention. In contrast, four years later Cruikshank (Figure 3.3) heralds Jenner as the saviour of children with smallpox, while doctors who oppose the procedure are running away surrounded by the corpses of those they did not vaccinate; a societal change of belief related to a preventative measure in four years (if these cartoons are a true reflection of public opinion).

What happened? One interpretation is that celebrity culture and the importance of opinion leaders was alive and well in the early nineteenth century. King George III had his family vaccinated and this was public knowledge. The discovery of vaccination was one thing but how to make it available to a wider population was entirely another. The first problem was that cowpox was and is an uncommon disease of cattle. Therefore, to

FIGURE 3.2 Images of Vaccination (1802 Gillray cartoon). The Cow-Pock or the Wonderful Effects of the New Inoculation!
Produced for the Anti-Vaccination Society. en.wikipedia.org/wiki/File:The_cow_pock.jpg

source an effective supply of vaccinia was extremely difficult. Second, what was the most effective means of delivery? How reproducible was the method, how many lesions, when and what were the correlates of protection? These questions are posed in the terminology of today, but no methodology then existed to measure effectiveness. Not even the rudimentary 'body bag count' of the vaccinated vs. the unvaccinated after exposure to smallpox was in use.

The supply problem was dealt with by arm-to-arm vaccination from one individual to another. There were side effects linked to poor technique and contamination. This was a particular issue in London because vaccination was often conducted in smallpox hospitals! It meant that when the fluid in a vaccinia lesion (lymph) was taken it was unclear whether this was vaccinia or the subject was incubating smallpox, paradoxically promoting spread of smallpox rather than protecting against it.

FIGURE 3.3 VACCINATION against SMALL POX. Mercenary & Merciless spreaders of Death & Destruction driven out of Society. Isaac Cruikshank (1808) Wellcome Images

How long did the protection last? Here Jenner was misrepresented. As early as 1819 he was discussing the need for re-vaccination to sustain protection. Re-vaccination was widely practised on the continent but less systematically in the United Kingdom.

With all these fundamental problems it is of little surprise that early opposition and scepticism to vaccination arose. This repeats itself with every health measure, especially when a public health intervention is proposed or introduced. It is even more of an issue today because of rapid but often ill-informed communication. Therefore, opposition follows like night follows day, so why are we always surprised when it does happen?

What were the main themes underlying opposition to vaccination? The first returns to the Gillray cartoon (Figure 3.2); introducing material from cattle was considered akin to bestiality. This general opposition to using animal material in vaccinations is exemplified by Dr Massey, who was an influential anti-vaccinator. He reported that he saw the case of an 'ox-faced boy', and attributed the transformation of the boy's face in 'assuming the visage of a cow' to vaccination. Anecdotes and opinion from so-called authorities are very powerful and still disproportionately influence medical and public health decisions today.

Arm-to-arm vaccination was the only means of effective early delivery, and this presented the problem of who you chose to collect material from for vaccination. The active nature of the material was unknown in a society dominated by social status, and selection of the donor was important to certain strata of society. The donor child should be from 'a good family' because it was wrongly assumed that the material used for vaccination would be of better quality and by implication safer. Again there are resonances with public health challenges today.

Perhaps more widely held were religious objections. Dr William Rowley stated in a sermon: 'Smallpox is a visitation from God originated in man but the cowpox is produced by presumptuous impious man, the former heaven ordained, the latter is perhaps a daring and profane violation of our religion.'[2] Roughly that translates into 'let them die but at least they'll go to heaven'. To counter such prejudice is difficult in a society that gives disproportionate credence to such statements, although today's parallels in other societies to the polio eradication campaign have resonance with this theme.

To counter such arguments a strong evidence base is essential, particularly as implementation of vaccination on a larger scale was not as effective as it could have been. Poor techniques, inadequate training and poor sourcing of the material, all contributed to the probability that a substantial fraction of vaccinations, even as many as half, were ineffective. But the implementation of this technology coupled with the monopoly exercised by the medical profession was starting to earn them a lot of money. Therefore, the pressure began to grow, from the medical profession and leaders who were convinced of effectiveness, to introduce compulsory uptake of vaccinations.

Between 1820 and 1840 the movement towards compulsion started on the continent of Europe. However, the approach adopted varied between countries. Each approach reflected the prevailing social attitudes in each country. In Prussia it was direct compulsion but in other countries it was more indirect, resulting in a 'Catch 22' proposition. For example, in Bavaria you could not get married without a vaccination certificate. This approach towards an artificial 'freedom to choose' again resonates with attitudes to childhood vaccination programmes today.

The movement to compulsion in the United Kingdom was gradual. In 1840, it was announced that vaccination would now be free under the provisions of the Poor Law Act, but this provision was accompanied, in 1853, by an Act of Parliament, using poor-law guardians, the Privy Council and the Registrar General to enforce universal vaccination. This proved an abject system failure. Simplifying this very complex system of multiple jurisdictions was difficult. It was supposed to work with the Poor Law Boards, through the Union Guardians, picking up responsibility for dividing the United Kingdom into districts, to which they were required to appoint competent 'medical vaccinators' and to provide for the location and timing to enable parents to bring their children to be vaccinated (Figure 3.4).

But competence of practitioners was under the supervision of the Privy Council. Parents had to notify the Registrar General (through local sub-registrars) of all births, as compulsory registration of births and deaths had been introduced, and on registration the parents received a notification

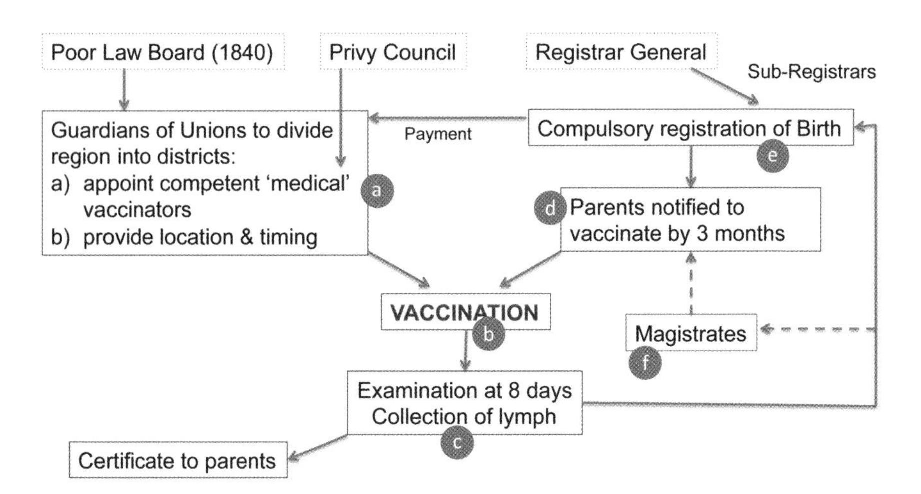

FIGURE 3.4 Implementation of the Vaccine Act 1853. Simplified schema for the introduction of compulsory vaccination in England and Wales. The main weaknesses of the system that led to inefficient uptake are indicated.

to have their children vaccinated. They would then go to the vaccination centre and return eight days after the procedure so that lymph could be collected from the lesion and a certificate of vaccination issued. This was then returned to the sub-registrar, who would ensure payment to the practitioner. Superficially, this system appears to have feedback and certification but it failed for several reasons:

(1) The training of 'medical' vaccinators was not uniform (a) nor was there an established 'standard operating procedure' (b) that could be implemented nationally. Therefore, training and vaccination (c) could be considered at best ad hoc.
(2) Numbers of 'vaccinators' were variable and not centrally determined, some keeping numbers small to maximise individual income.
(3) Notification and location of vaccination centres was variable; how was a parent to know when and where to go? Distance and how many centres in a 'district' were at the discretion of the guardians and not defined.
(4) Parents were suspicious of the process and did not believe in its effectiveness (d).
(5) The vaccinator determined 'vaccine take'; the individual who would be paid as a result of successful take was the very individual issuing a certificate. If they deemed the procedure a failure they would have to repeat it at their expense!
(6) If there was a failure of birth registration then there would be no payment and sub-registrars were neither well recompensed nor rewarded for efficiency (e).
(7) Failure of notification of vaccination was to be dealt with by magistrates fining parents (f). In many instances they were not prepared to do this.

Ultimately, there was no trust in the system, but parents were now in a position that they would be fined if they did not have their child vaccinated within three months of birth after compulsory birth registration. However, this has to be seen in the context of even more Draconian measures on the continent of Europe. The fix in the United Kingdom was to modify rather than overhaul the whole system with Acts in 1858, 1867 and 1871, each introducing tougher measures to force vaccination onto families.

The ultimate result of this compulsion was the emergence of anti-vaccination movements in Britain. These opinion formers and leaders coalesced their varying opposition to vaccination by uniting against compulsion. Examples include:

(1) Charles Creighton, the author of one of the most important treatises on epidemics and infection in the late nineteenth century. His view of vaccination: 'It is difficult to conceive what will be the excuse made for a century of cowpoxing, it cannot be doubted that the practice will appear in an absurd light to the common sense of the 20th century as blood-letting does to us now.'[3]

(2) William Tebb, a supporter of Robert Owen, and of many social causes (some unusual). He was founder of the Vegetarian Society and a prominent member of the Swedenborgian sect, an anti-vivisectionist, a spiritualist, theosophist and pacifist. He created the London society that fought to abolish compulsory vaccination, but ultimately his major achievement was the Society for Prevention of Cruelty to Children. He started a journal, 'The Vaccination Inquirer', which was geared towards opposition to vaccination. Today he is remembered for the bells located among the graves and mausoleums found in London cemeteries. His ultimate 'cause' was a society for the prevention of premature burial. It is noteworthy that in his will he stipulated he must lie for fourteen days till there were obvious signs of putrefaction before he was actually buried.

(3) Alfred Russell Wallace, the co-discoverer of evolution with Charles Darwin. His view of vaccination: 'Vaccination is a gigantic delusion, it has never saved a single life, it has been the cause of so much disease, so many deaths, utterly needless and altogether undeserved suffering. It will be one of the greatest errors of an ignorant and prejudiced age and its penal enforcement the foulest blot on a beneficent course of legislation during our century.'[4] His arguments with the Royal Commission to investigate vaccination are still the most lucid and clear attempt at an objective analysis of effectiveness.

(4) Lastly, if you thought it all went away in the nineteenth century, George Bernard Shaw writing in 1944 said: 'At present intelligent people do not have their children vaccinated and nor does the law now compel them to.... prophesied the extermination of smallpox but in contrary more people are now killed by vaccination than by smallpox.'[5] With respect to Smallpox he was right.

What did they have in common? They were all by today's political standards socialist and at least committed social reformers. Their actions were motivated out of a general sense to improve the lives of those in greatest need. But there are other odd associations into which we should probably not read too much: they were spiritualists; many linked to Swedenborgianism, vegetarianism, anti-vivisection and promotion of animal welfare. None of that is meant to be critical but it gives a flavour of how individuals who were engaged in addressing the ills of nineteenth-century society thought that vaccination was an important issue to engage with.

It is not surprising that compulsion was the over-riding and unifying issue, whatever people's personal views concerning vaccination. Accepted in Prussia, accepted on the continent of Europe, to an extent accepted in the United States, in Britain this interference with the rights of the individual by the state and 'authority' was the final straw. Financial penalties on those least able to afford them were cumulative, and anti-vaccination societies developed in virtually every city in Britain to oppose vaccination. But being opposed to vaccination was not seen in the late nineteenth century as being opposed to science. In fact some individuals argued that you were a more rigorous scientist if you explored whether vaccination was effective or not.

Their pressure resulted in a Royal Commission, which sat from 1888 to 1896. Its deliberations were published in many volumes of debate and submissions but finally summarised in the British Medical Journal in 1896. The promoters of vaccination in the medical profession were very strongly opinionated. John Simon wrote in a letter to the President of the General Board of Health: 'It goes with the credulity which characterises the present age to be incredulous of proved truth. Alike in rejecting what is known and believing what is preposterous, the rights of private foolishness assert themselves. It is but the same impotence of judgement which shrinks from embracing what is real and lavishes itself upon clouds of fiction.'[6] Many clinicians have often felt that way about medical issues even if they express it less eloquently than that.

But the promoters did not submit anything resembling the evidence-based argument we would expect today. Largely, they provided assertions, anecdotes and opinions. Contrast this with the work submitted by Alfred Wallace. He collected data; he tried to show the vaccination rates in cities in the United Kingdom, linking these to smallpox outbreaks that occurred. These data, particularly from Leicester, if examined superficially could be interpreted as suggesting that vaccination does not work.

But there was evidence available that vaccination was effective. One of the best examples comes from Sweden towards the end of the nineteenth century. Because of religious objection, parents in Stockholm stopped vaccinating their children. However, the rest of Sweden continued with vaccination. When the epidemic of 1873 hit, it was Stockholm that was affected, and on re-starting vaccination in Stockholm the epidemic went away. The rest of Sweden was marginally affected.

The decision of the Commission was to continue with the vaccination programme, but ultimately the objectors had their way with regards to compulsion. In 1906 conscientious objection to vaccination was allowed. In a sense it was a very British way of dealing with the problem: you can object to it conscientiously but actually we will keep the law there and only repeal it when it has fallen into disuse.

By 1966 the number of smallpox cases was diminishing globally. The WHO declared the world free of smallpox in 1980; a real success for public health and immunisation. But tragically, the last fatality was due to a laboratory accident in Birmingham. The debate now concerns whether the last remaining stocks of smallpox virus held in Russia and the United States should be destroyed.

So Alfred Russell Wallace was right and wrong in equal measure. He contributed enormously to our understanding of evolution and was one of the truly great Victorians. He stood for rigorous analysis and the right of parents and individuals to choose, and opposed centrally ordained compulsion, a very important societal principle; in this he was able to claim victory. However, he was wrong in claiming that his defining work to stop vaccination was his most important.

And the reason for this error in such a great thinker and scientist, I believe, was probably that he lost his objectivity. The only solution he saw consequent on his studies was the abolition of vaccination. If his data to the Royal Commission is re-examined, it could equally well be explained if vaccination was only partially effective. This is credible, bearing in mind how vaccination was conducted. As the system improved then the original observation made by Jenner stood vindicated.

Wallace went to his grave believing that he was right and he wrote in his own biography that 'I feel sure that the time is not far distant when this will be held to be one of my most truly scientific of my works'. He thought this was more important than the theory of evolution. He was wrong; it was not his best work.

Introducing National Vaccination Programmes

So what have we learnt from this episode about the introduction of public health measures to prevent plagues? In the twentieth century we learned lots about the underlying principles of vaccination and the correlates of protection, and we have a greater understanding of immunology to explain why these interventions work. Modern approaches to the development of vaccines using all the opportunities afforded by molecular biology mean that we can develop vaccines against most pathogens quickly and effectively. There are some difficult organisms that present particular biological problems, such as HIV, TB and malaria, but vaccination/immunisation is one of the defining achievements of public health intervention in the twentieth century. Yet we have not grasped the fundamental message that is so well exemplified by the nineteenth-century opponents of vaccination. This concerns the very human nature of scepticism and opposition to acceptance of novel interventions often propounded by a very vocal and influential minority.

How did these issues play out with the introduction of the Measles, Mumps and Rubella (MMR) vaccine? The *Daily Mail*, who led much of the campaign against MMR, identified the importance of opinion leaders, such as the prime minister, failing to be open about whether their children had had the MMR vaccine. There is an important issue of privacy that the prime minister invoked but nevertheless this was not the case when vaccination was introduced. It is difficult to be certain of the impact of such individuals coming forward but certainly the tabloid thought this was an important event in the timeline of the MMR immunisation story.

The importance of media in developing the opposition to MMR was underlined by Tom Whipple writing in *The Times*. When the first stories concerning a possible association between MMR and autism emerged in 1998/99 those stories were uniformly critical of the MMR vaccine. The vaccination rates for MMR started falling. Despite further coverage, with many more articles in the popular press concerning MMR, many being critical of the association previously described and much more supportive of the immunisation, this had little impact in the plummeting rates of MMR immunisation. Fascinatingly, correcting the original errors did not assuage the public nor did it serve to reverse the downward trend of measles immunisation. His overarching conclusion was that the press could not undo what they started; therefore the best thing journalists could do about MMR is 'shut up'.

However, the predictable consequences were all too evident last year, when in the Swansea region of South Wales, measles cases rose dramatically. The population had a low level of 'herd immunity' due to low immunisation rates and was susceptible to an outbreak when exposed to the virus.[7] Therefore, in today's world, where information flow is so rapid, the media is an important vehicle for education. They are themselves key opinion leaders, producing potentially dramatic effects both for good and ill.

Measles is an important case in point in the global perspective. While measles vaccines are very effective, safe and relatively inexpensive, this infection is still a leading cause of death amongst children worldwide. The most recent WHO figures (2013) reveal that there were more than 145,000 deaths that year, equalling approximately 16 deaths per hour worldwide. A close relative of measles, rinderpest (shared a common origin with measles in the tenth or eleventh century), which was the cause of a serious and often fatal disease of livestock, has been eradicated by veterinarians through an effective global vaccination programme. Given the global access to measles vaccines, this infection, like rinderpest, could be eradicated if minority public opinion or belief could be addressed. A Californian vaccine bill enforcing vaccination of all school children has been passed following a significant measles outbreak in that state. Will the imposition of mandatory vaccination over parental choice or religious belief help eradicate measles? The 2015 Californian vaccine bill is sure to fuel this debate.

So the evidence would suggest that lessons from history that should have been learnt are not, but as a society we keep repeating the same mistakes, especially in our interaction with the public at large. We keep assuming that society accepts the prevailing biomedical construct, but this is just not the case, and objection and opposition may not be rational in terms of prevailing evidence. Yet it must be taken seriously. History teaches us that there will always be opponents to any proposed measure, and in an information rich and communication savvy era all the more effort needs to be expended in debate. Sometimes that debate concerns the age-old issue that has surrounded all public health intervention from the time of the Black Death through to MMR – the right of the individual over the needs of society. As more and more complex preventative measures are proposed, most now require active participation of the individual, who may make different decisions concerning their personal perception of the risk-to-benefit ratio.

New Opportunities in Plague Prevention

Are we getting better in promoting this debate? Two examples of proposed measures that could impact on all of us are cervical cancer and the prevention of influenza pandemics. The cervical cancer vaccine, widely used today, is an intervention that could save the lives of up to 250,000 women who still die of what is the second most common cancer in women globally. Trials of the vaccine show protection from the human papillomavirus that causes cervical cancer for ten years and it is effective in clinical trials in reducing the early signs of the disease. All clinical evidence points to this being a worthwhile public health intervention that is even more important in the poorer parts of the world than in developed countries in North America and Europe.

However, the immunisation rates for cervical cancer vaccines vary considerably. Uptake in Britain has started well but in the United States there is no national approach and they are reliant on individual participation, which is barely hitting 30 per cent. Why? Numerous explanations are proposed but many focus on the negative publicity consequent on the belief that this immunisation is being promoted by industry. This alone is too simplistic an explanation, but current data suggests that a concerted national campaign may have far greater impact on uptake.

And the media continue with headlines such as those of November 2011, 'Girl left in waking coma as a severe reaction to cervical cancer jabs'. The following day 'cervical cancer jab left girl, 13, in waking coma'.[8] It is interesting that now these stories are challenged and these headlines were amended after Sense about Science objected to the Press Complaints Commission. The Press Complaints Commission upheld the judgement and expected that greater care would be taken when presenting articles on health issues in the future.

That is progress. It is slow and many of us might wish to see it accelerated but we must always ensure balance and engage in the debates that surround these issues otherwise those engaged in developing such interventions will be falling into the trap of the proselytisers of the nineteenth century.

The Acceptability of New Approaches to Protection from Plagues

In research there are singular events that impact disproportionately on personal perception. For me, one such was the book written by Roy Anderson and Robert May, *Infectious Diseases of Humans*. They showed that even rudimentary modelling techniques could predict the behaviour of epidemics due to infectious agents both in man and other species. During my time at the MRC, I was particularly pleased that we started the Centre for Outbreak Control, headed by Neil Ferguson and based at Imperial College.

As a non-mathematician, my understanding relates to the practical utilisation of these novel approaches in a real-life setting. Suffice to say that these models work on the basis of a point of infection occurring in a population which can be considered in three distinct phases relative to the pathogen: susceptible, infected and recovered. In their analysis they showed that two parameters were of major significance in determining outcomes of epidemic spread, R_0 and T_G. However, this is difficult data to collect; yet their conclusions make it essential that we make a concerted effort to do so. Thus modelling changes the way in which an investigator should approach trying to understand the biology of pathogens.

The first of these parameters is R_0, which is the basic reproduction number. This is the number of secondary cases generated by a single primary case entering a susceptible population. So, that number defines the intensity of an outbreak. Second both R_0 and T_G, (the generation time; the average time for secondary cases to be infected by the primary case) will help identify how much time is available to intervene on a population-wide basis.

This simple starting model requires a variety of modifications in a real-world setting. In keeping with the theme of objectivity, what is the evidence that this approach can predict reality? The UK foot-and-mouth disease outbreak in cloven-hoofed animals in 2001 was modelled retrospectively by Keeling and May[9] to 'predict' where the outbreak would be located, and its intensity, based on movement of animals, farms and susceptibility to the virus. The results are uncanny in their predictive ability and the methods could have been used to model the outbreak accurately and facilitate effective intervention.

As the models become more and more sophisticated, we can begin to use them to develop interventions in the event of human epidemics. To most investigators one of the most worrying possibilities is the occurrence of an influenza epidemic. This virus has a particular propensity to mutate because of the nature of segmented RNA viruses.[10] Such mutations can be enhanced by intermediate hosts, birds and pigs in the case of influenza, and can then transmit to humans, where case-to-case spread occurs rapidly. Historically, such outbreaks have caused devastation. A new strain of influenza in 1918 (Spanish Flu) killed probably more than 100 million people, more than all deaths in World War One.[11] Subsequent major changes in the virus resulted in approximately three pandemics in the preceding century. Therefore, we know that another pandemic is inevitable. However, we cannot predict where and when it may strike or its likely severity (the last in 2009 was relatively mild in its global impact). This is a real challenge and we must run these models based on the best available evidence.

Ferguson and colleagues have studied factors that can be used to contain a flu outbreak.[12] Two short video clips[12] are available, which model a period of sixty days from the onset of a point outbreak of ten cases commencing in a rural area of Thailand. His team utilised models of population movement in the country as a major factor in the network through which this infection spreads.

As simulated transmission models run on large computers demonstrate, a small initial outbreak would rapidly extend, as quickly as thirty days after the onset of detected infection. Recovery, with consequent resistance to infection, or mortality also spread gradually in the affected areas. With basic models such as this, they were able to ask questions such as, what if different control interventions were instituted? Which should be explored? Mass immunisation will be difficult to institute on this timescale, but an effective prophylactic therapeutic agent could be. The impact of such an intervention is examined in a clip that can be viewed online.[12] Modelling suggests that an outbreak can initially be brought under control by the use of a therapeutic agent. But, if the infection crosses a border, it spreads and ultimately is reintroduced and re-invades as the drug supply to exercise the initial control is exhausted. From such studies, Ferguson and colleagues drew important conclusions relating to what a global response might look like. Variation in policies between countries is liable to make the situation worse rather than better. Most striking of their conclusions was that to localise an outbreak and therefore to prevent a regional and global pandemic can only be effective if the first forty cases are detected. This is a difficult proposition in the parts of the world where an epidemic is likely to occur. It follows that any response has to be international from day one in order to control an international pandemic that could result. These conclusions are based on parameters and assumptions that are set very low for transmissibility. If a mutation in the virus were to change this significantly then the result is far worse.[12]

The control of plagues has moved now from developing interventions that impact on a single country to a situation triggered by the advent of modern travel and migration, which needs interventions that cross boundaries. It is difficult enough to gain acceptance to restrictive measures in one society but extending this to multiple societies with their consent and agreement is a real challenge to diplomacy and international collaboration. The 2014 Ebola outbreak in West Africa is a contemporary case in point (Chapter 1).

Natural plagues are difficult enough but the scenarios are further changed with the deliberate creation of a plague as an act of terrorism, where the nature of the infectious agent can be modified for maximum devastation. Riley and Ferguson have investigated this using smallpox as a potentially reintroduced virus.[13] Many of the parameters for this virus are known from multiple studies and these assumptions define the profile of the epidemic; they include the infectious period and the subsets of individuals who actually will be infected. But it has to be remembered that these can be varied by manipulation of the genome of the agent concerned.

Deliberate release in the United Kingdom was considered as a point outbreak in Central London. It is mapped using data of population movement occurring at a 5 km radius throughout the United Kingdom. Using the 'normal' parameters for smallpox such an outbreak can be controlled pretty much within London. If this control is exerted, about seventy-five days are available to vaccinate the remaining susceptible population of the whole country, provided new batches of vaccine can be manufactured. In this setting, standard control measures would work. But to prevent a man-made catastrophe with deliberately introduced variation into the infectious agent is another matter. Modelling and scenario planning are improving, and these will increasingly play an important role in future control of infectious plagues. Perhaps the most difficult debate for the future will be to gain acceptance that any planning that restricts the rights of individuals in any society is indeed acceptable in the wider community.

The Plague of Non-Infectious Diseases

Plagues are still widely considered as being caused by infectious agents. However, in the modern world, increased longevity and unhealthy lifestyles mean that future, and indeed current, epidemics are those of chronic non-infectious disease such as cardiovascular, respiratory diseases, diabetes and cancer. Projected deaths and burden of disease in India indicate that globally we have already moved away from infectious diseases as a major cause of mortality and these modern plagues are taking over.[14] This is a major topic in itself and it is impossible to consider all the ramifications for public health and policy. But we are not helpless in relation to intervention, although many such interventions will not be classical therapeutic approaches. Instead they will rely on all sectors of society, commercial interests and government working in tandem. Therefore, we have a different scenario, one that will require even greater engagement, acceptance and participation by individuals not just for their own direct benefit but also for the benefit of others. Openness and transparency will never be more important than in these debates.

One example may illustrate this point. As China is getting richer, the production of cigarettes by China has grown such that it produces 42 per cent of the world's cigarettes, a large source of tax revenue. Moreover, China now has 350 million smokers. Numbers of cigarettes consumed per day in 1952, 1972 and 1992 rising to a national average of ~10/day resembling data from 1910, 1930 and 1950 when it reached similar levels in the United States. The consequence was that 12 per cent of all deaths in America in 1950 were attributable to tobacco, rising to 33 per cent in 1990. If this is paralleled in China then a serious epidemic will arise because of continued use of tobacco. However, the state can intervene effectively through a variety of measures. Some are already being introduced,[15] but will one of the most effective, pricing, be used? In Europe, cigarette consumption in France rose into the 1990s, but increasing taxation and prices meant that more people stopped smoking. So the state can intervene and make a difference but it has to balance the public acceptability of such measures and their impact on other sectors of the economy. But it is evident that future interventions to control these modern plagues are going to have a large component of health education.

Conclusion

In today's noisy world, major public health issues abound as communication, debate and information concerning analysis of association and causality increase. This is paralleled by our greatly increased capacity to intervene and monitor effectiveness of implementation. We have better predictive tools and measures and these are going to continue to improve. Therefore, I believe the real challenge is going to be the acceptability of intervention by society at large as a key to the future prevention of plagues of all types. Will we accept the concept enshrined in the common paraphrasing of Jeremy Bentham's 'greatest good for the greatest number' rather than his original use of 'happiness' rather than 'good'?[16]

However, appealing as this may be, we have to recognise that such interventions will always produce individual casualties and a significant restriction on the rights that individuals enjoy in a free society. In the face of overwhelming threat, that decision is superficially easy but the moral and ethical dilemma becomes more focussed if we utilise predictive measures of risk that result in measures that constrain our actions.

What is the impact on individual rights? The age-old question of the rights of society versus the rights of the individual remains: who to judge and who to decide? This debate will always be coloured by the nature of an individual's belief in the value of an intervention and/or their perception of risk to them and society. As the history of plagues teaches us, never make the assumption that an individual's reasoning is going to be rational and conforms to prevailing 'expert opinion'.

What does it mean for places of study such as the University of Cambridge? Well, simply that understanding how society and individuals behave in the face of the threat of a plague and consequent preventative intervention is a hugely complex field for study. In the case of biological and health-related threats it goes far beyond the biological explanation or the biotechnical aspects of developing an intervention. But it is an exciting area for interdisciplinary study that can engage all academics. Or put more prosaically, for those wedded to disciplinary boundaries, sociology, mathematics and policy studies are as important as biology to help successfully prevent plagues.

References

1. The Epidemiology of Small Pox: http://apps.who.int/iris/handle/10665/39485.
2. https://archive.org/stream/b21536880/b21536880_djvu.txt.
3. http://whale.to/vaccines/creighton_b.html.
4. Wallace: https://piotrbein.files.wordpress.com/2014/01/vaccination-a-delusion-by-a-r-wallace-1898.pdf.
5. George Bernard Shaw: https://therefusers.com/intelligent-people-dont-vaccinate-their-kids-george-bernard-shaw-nobel-prize-winner/.
6. Quain, R. The Harveian Oration on the history and progress of medicine. *BMJ.* 1885; **2**: 775–781.
7. Outbreak of Measles in Wales Nov 2012–July 2013. *Report of the Agencies which Responded to the Outbreak.* (2013) www.his.org.uk/files/9213/8442/3157/Final_measles_Oct_Report.pdf.
8. https://childhealthsafety.wordpress.com/2011/11/17/girl-13-left-in-waking-coma-by-cervical-cancer-jabs/.

9. Keeling M.J., Woolhouse M.E.J., May R.M., Davies G., Grenfell, B.T. (2003) Modelling vaccination strategies against foot-and-mouth disease. *Nature* **421**:136–42.
10. Webster R.G. (2013) *Textbook of Influenza.* Wiley; 2nd edition.
11. Barry J.M. (2009) *The Great Influenza: The Story of the Deadliest Pandemic in History.* Penguin Books; Revised edition.
12. Ferguson N.M., Cummings D.A.T., Cauchemez S. *et al.* (2005) Strategies for containing an emerging influenza pandemic in Southeast Asia. *Nature* **437**:209–14. www.nature.com/nature/journal/v437/n7056/extref/nature04017-s4.mov.
13. Riley, S., Ferguson, N.M. (2006) Smallpox transmission and control: Spatial dynamics in Great Britain. *PNAS* **103**:12637–42.
14. Daar A., Singer P., Prasad D. *et al.* (2007) Grand challenges in chronic non-communicable diseases. *Nature* **450**:494–6.
15. Levy D., Rodríguez-Buño R.L., Teh-Wei Hu, Moran A.E. (2014) The potential effects of tobacco control in China: Projections from the China SimSmoke simulation model. *BMJ* **348**:g1134.
16. www.friesian.com/bentham.htm (no. 3).

Further Reading

Ziegler P. (1982) *The Black Death.* Penguin Books UK.

McNeill W.H. (1977) *Plagues and Peoples.* Anchor Books.

Kelly J. (2006) *The Great Mortality: An Intimate History of the Black Death, the Most Devastating Plague of All Time.* Harper Perennial; Reprint edition (31 January 2006).

European Centre for Disease Prevention and Control (2014) *Outbreak of Ebola Virus Disease in West Africa.* Stockholm: ECDC.

Callaway E. (2011) Plague genome: The Black Death decoded. *Nature* **478**:444–6.

Clifford J.G. (1989) Eyam Plague, *1665–1666.* Eyam: J.G. Clifford.

Lehohla P.J. (2013) Mid-year population estimates 2013. Statistical release P0302. http://beta2.statssa.gov.za/publications/P0302/P03022013.pdf.

Williams G. (2011) *Angel of Death: The Story of Smallpox.* Palgrave Macmillan.

Boylston A.W. (2012) *Defying Providence: Smallpox and the Forgotten 18th Century Medical Revolution.* CreateSpace Independent Publishing Platform.

Ferguson N., Keeling M.J., Edmunds W.J. *et al.* (2003) Planning for smallpox outbreaks. *Nature* **425**:681–5.

Wilkinson S. (2007) *The Vaccination Controversy*. Liverpool University Press.

Creighton C (1887) *The Natural History of Cowpox and Vaccinial Syphilis*. Cassell & Co.

Wikipedia – William Tebb (accessed 2014) http://en.wikipedia.org/wiki/William_Tebb.

Smith C.H., Beccaloni G. (eds) (2008) *Natural Selection & Beyond; the Intellectual Legacy of Alfred Russel Wallace*. Oxford University Press.

Wallace A.R. (1908) *My Life: A record of events and opinion*. Cosimo Classics, New York.

The Report of the Royal Commission on Vaccination (1896) *BMJ* **2**:453–9.

Borysiewicz L.K. (2010) Prevention is better than cure. *Lancet* **375**:513–23.

Whipple T. (2009) Newspapers must shut up about MMR. *The Times* (London). June 22, 2009. www.timesonline.co.uk/tol/life_and_style/health/child_health/article6549259.ece.

Lehtinen M., Dillner J. (2013) Clinical Trials of human papillomavirus vaccines and beyond. *Nature Rev Clin Oncol.* **10**:400–10.

Human Papillomavirus (2012) *Nature* **488**:S1–S48.

Sense about Science (2011) www.senseaboutscience.org/for_the_record.php/77/quotcervical-cancer-vaccine-has-left-our-daughter-in-waking-comaquot.

Sense about Science (2012) *Memorandum: Leveson Inquiry: Culture, Practice and Ethics of the Press*. www.levesoninquiry.org.uk/wp-content/uploads/2012/02/Submission-by-Sense-about-Science.pdf.

Anderson R., May R. (1992) *Infectious Diseases of Humans: Dynamics and Control*. Oxford University Press.

The Economist (2014) Government coughers: Smoking is on course to kill 100 m Chinese people this century. Will the latest anti-smoking policies curb it? www.economist.com/news/china/21597958-smoking-course-kill-100m-chinese-people-century-will-latest-anti-smoking.

4 The Nature of Plagues 2013–14

A Year of Living Dangerously

ANGELA MCLEAN

The 12 months from June 2013 to May 2014 were, in many ways, typical in the emerging infectious disease events that occurred. There were no huge shocks, no massive outbreaks nor new pandemics, but every month there were important events and together the year's events form a good illustration of what is a 'normal' rhythm of events for emerging infectious diseases. However, after May 2014 the Ebola epidemic in West Africa (described, in its infancy, under 'March' in this chapter) rapidly expanded to become a very large epidemic, illustrating how quickly small outbreaks can become very large problems given circumstances that favour human to human transmission and rapid spread.

Whilst many people think of 'emerging infections' as only the brand new infections like SARS and HIV, the definition of emerging infections is broader and includes five types of infections that are in some sense 'new'. Table 4.1 describes those five types and gives examples of each from the past.

In England, Public Health England (an agency of the Department of Health) routinely gathers up evidence about new infectious disease both nationally and internationally. This 'horizon scanning' activity is an important part of identifying new infectious hazards that may pose a risk to public health. Each month Public Health England, along with other government bodies, publishes a two-page summary of notable events of public health significance.[1] These summaries are widely circulated in government and academia and are publically available. They form both an excellent warning of current events and a record of how events unfold over months and years.

In this article I have picked one event from each of the past twelve months to illustrate the 'normal' rhythm of incidents. Those events have been chosen to illustrate the five types of emerging infectious disease events. They include the three events of 2013–14 that are most likely to

Table 4.1 *Five types of emerging infectious disease events.*

Type of event	Historical example
A novel infectious pathogen never previously seen or not previously seen in man	HIV (1983) and SARS (2002)
A pathogen spreading into new geographical areas	West Nile Virus spreads to N. America (1999)
The re-emergence of an infectious agent that was previously being brought under control	Tuberculosis (ongoing)
Newly evolved variants of known infections	Influenza and drug-resistant malaria (ongoing)
A pathogen newly associated with a known disease	*Helicobacter pylori* and stomach ulcers (1983)

trigger substantial, global problems in the future: the ongoing MERS-coronavirus outbreak in the Middle East (July 2013), the ongoing zoonotic cases of Avian Influenza in China (February 2014) and the re-emergence of Polio in early 2014 (May 2014). Despite the ongoing fears about a devastating influenza pandemic, the biggest realised threat from emerging infections continues to be the evolution of antimicrobial resistance. This is a slow, chronic problem that is happening everywhere all the time and therefore never triggers a single 'event'. However, during these twelve months WHO published its first global report on surveillance for antimicrobial resistance.[2] The findings from this report are described here under April 2014 in recognition of the importance of antimicrobial resistance as a form of emerging infection.

The article commences with a map showing where the twelve incidents occurred. Each incident is then described in turn before the article ends with a discussion of what we can learn from studying these dozen emerging infectious disease events (see Figure 4.1).

June 2013: a Novel Cyclovirus

The year in question kicked off with the discovery of a new virus: found in the cerebrospinal fluid of twenty-six patients with acute central nervous

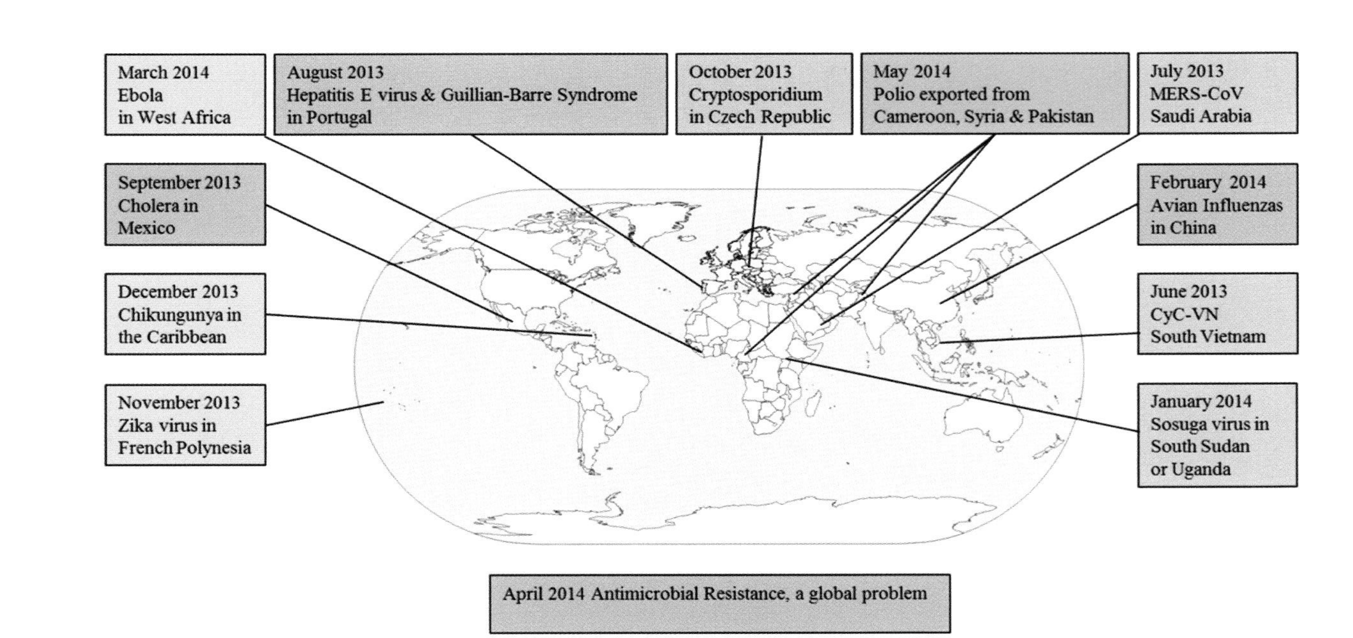

FIGURE 4.1 Twelve emerging infectious disease events around the world: June 2013–May 2014. The colour coding follows the same scheme as Table 4.1. In blue are the novel infections, in green infections spreading to new geographical areas, turquoise denotes a re-emergence event, pink a newly evolved variant of a known infection and violet a new association between a known infection and a known disease.

system infections from Southern and Central Vietnam.[3] This new virus belongs to a family of viruses (the Cycloviruses) that infect many species: genomes have been detected in samples from humans, a range of other vertebrates and in insects.[4]

The new findings from Vietnam were interesting because they raised the possibility that this new virus was a zoonotic cause of acute central nervous system infections. Acute infections of the central nervous system are a substantial cause of morbidity and mortality but in the majority of cases it is not possible to identify the causative agent. Finding novel pathogens that cause such diseases is therefore important. The novel cyclovirus (named CyC-VN) was present in the cerebrospinal fluid of 4 per cent of patients with acute central nervous system infections whilst absent in the cerebrospinal fluid of control patients with non-infectious neurological disorders. It was also present in the faeces of healthy children, pigs and poultry from the same regions of Vietnam. However, no definitive causal link was claimed and the European Centre for Disease Prevention and Control warned that 'increased sensitivity may on occasion lead to spurious associations'[4] and that further studies were warranted to assess the risk posed by the newly identified virus (see Table 4.2). A follow-up study published six months later found only limited geographic distribution,[5] illustrating very clearly that sometimes what appears to be an exciting new finding is only of local significance.

It is not yet clear if CyC-VNvirus is causing disease in South and Central Vietnam, or is just associated with it. What is clear is that modern genomic techniques now allow the identification of infectious agents without the need to grow them in laboratory culture. These metagenomic techniques[6] have opened a window on a whole new world of microbiology that was previously unknown. This new-found ability to identify novel infectious agents will undoubtedly find important new pathogens, but will also create false leads. Separating the signal from the noise will be a major task.

July 2013: Middle East Respiratory Syndrome Coronavirus (MERS-Cov)

Middle East respiratory syndrome (MERS) is a new disease caused by a newly discovered virus called MERS coronavirus (MERS-CoV). It was

Table 4.2. *The prevalence of the newly discovered virus, a cyclovirus, designated CyCV-VN, in samples from a range of patients, healthy controls, animals and geographic locations.* [3,5]

Prevalence	%	Sample type	Patients	Location
10/273	3.7%	Cerebrospinal fluid	Adults and children with acute central nervous system infection of unknown cause	South and Central Vietnam
16/369	4.3%	Cerebrospinal fluid	Adults and children with acute central nervous system infection with other pathogens	
0/122	0%	Cerebrospinal fluid	Patients with non-infectious neurological disorders	
8/188	4.2%	Faeces	Healthy Children	
38/65	58%	Faeces	Pigs and Poultry	
0/615	0%	Cerebrospinal fluid	Acute central nervous system infections	North Vietnam, Cambodia, Nepal, Netherlands

first described in September 2012 and by the end of July 2013 there had been ninety-four laboratory-confirmed cases, of whom forty-six had died. Common symptoms are fever, cough, shortness of breath and muscle pain and many patients also have digestive tract problems. The great majority of cases are reported in Middle Eastern countries. It is thought that dromedary camels are the source of human infections, but human-to-human transmission is possible and is a particular problem in health care settings.

During July 2013, the growing number of cases and high case fatality rate caused the World Health Organisation (WHO) to convene an

Emergency Committee to advise whether MERS-CoV constituted a 'Public Health Emergency of International Concern'. The term 'Public Health Emergency of International Concern (PHEIC)' is defined in the International Health Regulations (2005) as:

> an extraordinary event which is determined...: (i) to constitute a public health risk to other States through the international spread of disease; and (ii) to potentially require a coordinated international response

The advice of the Emergency Committee was that the conditions for a PHEIC had not been met, but that the situation clearly warranted better surveillance, infection control, risk communication and research.[7]

In the ensuing ten months the number of cases increased dramatically. By the end of May 2014 there had been more than 665 laboratory-confirmed cases, including 205 deaths. Of these, about one-half were primary cases, and the other half were secondary cases (i.e. having had known contact with a confirmed case before they fell ill). Sources of infection thus fall into three groups: zoonotic, human-to-human in a health care setting and human-to-human in households. At the time of writing the dramatic increase in the number of cases is attributed to three possibilities: increasing zoonotic transmission, sub-optimal infection control in hospitals and better case detection.[8]

During May 2014 WHO's Emergency Committee was convened again. It noted its concern about the growing number of cases and indicated that 'the seriousness of the situation had increased in terms of public health impact'. However at that moment there was no evidence of sustained chains of human-to-human transmission and for this reason the committee determined that the conditions for a PHEIC had not been met.[9]

MERS-CoV carries its genetic information as RNA. RNA viruses have particularly high mutation rates and there is therefore concern that MERS-CoV may adapt to its new human hosts and become better able to transmit from one human to another.[10,11] At the moment it is estimated that although human-to-human transmission is possible, each case, on average, causes less than one secondary case. Under these circumstances sustained chains of transmission are not possible. The ongoing concern is that through viral adaptation this could change and that large epidemics or even a pandemic might ensue.

August 2013: Hepatitis E Virus as the Cause of Guillain-Barré Syndrome

August's story is a case report of an individual who developed Guillain-Barré syndrome triggered by infection with Hepatitis E virus.[12]

Guillain-Barré syndrome is an autoimmune disease of the peripheral nervous system. In about 60 per cent of cases Guillain-Barré syndrome occurs after a bout of infection with one of a number of bacteria or viruses. Hepatitis E virus is a cause of viral hepatitis. It is rare in wealthy countries and is thought to be most commonly acquired through consumption of undercooked meat products.

This case report highlights a relationship between a known virus and a known disease. This association had been described before,[13,14] so this report adds weight to an emerging view that this Hepatitis virus can trigger Guillain-Barré syndrome. Thus this is the type of emerging infection in which an old disease is newly associated with a known infectious agent.

September 2013: Cholera Re-Emerges in Mexico

In September 2013, after more than a decade of virtual absence, Cholera re-emerged in Mexico. By the time the outbreaks ended in mid-November there had been 180 confirmed cases and one death. This was the first time there had been sustained transmission of Cholera in Mexico since it had been brought under control in 2001 (Figure 4.2).

The strain of Cholera that caused the 2013 outbreak in Mexico was different from that which had circulated during the endemic period from 1991–2001, but similar to strains circulating in Haiti, Dominican Republic and Cuba at the same time.[16] All the evidence pointed towards multiple introductions to Mexico from neighbouring states which had been experiencing sustained Cholera epidemics since Cholera re-emerged in Haiti in 2010.[17]

October 2013: Hedgehog Associated Cryptosporidium, First Human Case

A spectacular example of the power of modern molecular diagnostics gave rise to the story from October. An immunocompetent man with

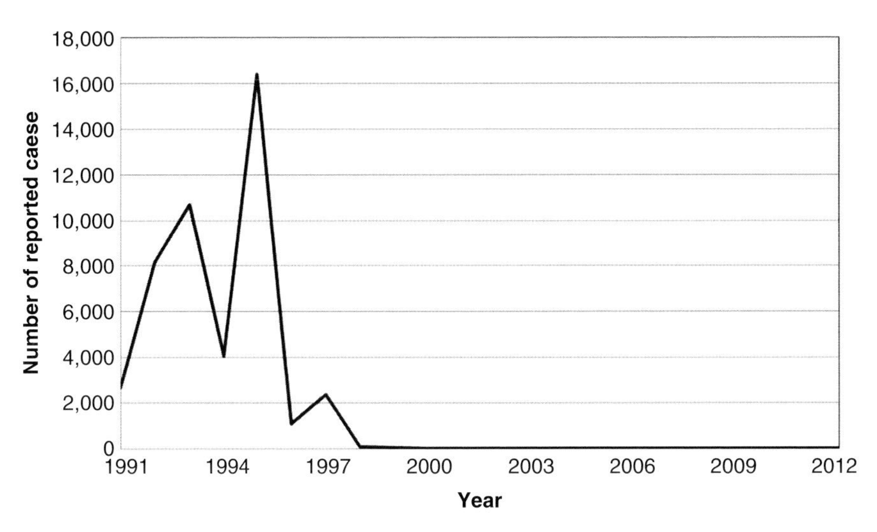

FIGURE 4.2 Cholera in Mexico 1991–2012[15]

gastroenteritis was found to be infected with a genotype of *Cryptosporidium* first described in hedgehogs.[18] The patient did not report recent hedgehog contact, and as the brief report so aptly put it 'further research is required to determine the transmission route'.

November 2013: Zika Virus Outbreak in French Polynesia

Zika virus is a mosquito-borne virus. It was first discovered in Uganda in 1947 and outside Africa and Asia is designated an emerging infectious disease. In November 2013 an outbreak was described in three archipelagos of French Polynesia. The symptoms of Zika virus infections are generally mild and it is considered to be a self-limiting febrile illness which lasts for about a week. The largest, previous, well-described outbreak was in Yap (Federated States of Micronesia) in 2007 and consisted of thirty-one cases.[19] In November 2013 the French Polynesian outbreak stood at 400 clinically suspected cases. However, by early February 2014, 8262 suspected cases had been reported through a syndromic surveillance network.[20] About half the samples sent for laboratory confirmation were confirmed to be Zika virus infections by genomic analysis. Furthermore, more than 28,000 individuals (more than 10 per cent of the population of French Polynesia) had sought medical care with Zika-like symptoms. With this huge burden of infection, uncommon complications became a problem, even though this is normally a mild, self-limiting disease. For example, during the outbreak there were thirty-eight cases of Guillain-Barré syndrome subsequent to Zika virus infection. This is four to ten-fold more than the usual annual number of cases of Guillain-Barré syndrome in French Polynesia and the complex clinical needs of these patients put a severe stress on intensive care resources in this very remote setting (ECDC 2014a).[8]

As an emerging infection spreads into a population without prior immunity very large outbreaks like this are possible if transmission is suitably efficient. Even though the great majority of Zika infections are mild and self-limiting, if enough people are infected, even rare complications start to impose a serious public health problem.

December 2013: Chikungunya Virus in the Americas

Chikungunya virus is another mosquito-borne infection. It has never before been known to be transmitted in the Americas. The symptoms are more serious than those of Zika virus infection, consisting of fever and joint pain which can last for weeks or months.[21] It is endemic in parts of Africa, SE Asia and India. There was a large outbreak in 2005–6 that started on Reunion[22] and smaller outbreaks in Italy and France in 2007 and 2010, respectively.[23,24]

In early December 2013, two cases of Chikungunya were reported on the French part of the island of St Martin.[25] The epidemic then took hold with cases doubling approximately every two weeks, so that six months later, by mid- June 2014, there had been over 160,000 suspected cases and 14 deaths with cases in 19 different Caribbean countries (see PAHO 2014[26] for updated figures). Infection spreads well in Caribbean islands because the population has never been exposed before, so there is virtually no acquired immunity and the mosquitos that can spread Chikungunya are present. These two risk factors are equally present in large parts of South and Central America, the South-Eastern parts of the USA and Southern Europe. The presence of very large Chikungunya outbreaks in the Caribbean increases the risk of very large outbreaks elsewhere in the Americas, particularly in poorer countries in Central and South America where control of the transmitting insect is more difficult.

January 2014: Sosuga Virus, a Novel Virus in a Wildlife Biologist

January 2014 saw the discovery of Sosuga virus, a novel virus isolated from a hospitalised wildlife biologist.[27] In 2012 a wildlife biologist fell seriously ill shortly after returning to the USA from a six-week field expedition to South Sudan and Uganda (hence the name of the new virus, Sosuga). Five days after returning to the USA she was admitted to hospital with a fever. Because of her work she had been in contact with a large range of bats and rodents and this raised the possibility that her illness was one of the very serious viral haemorrhagic fevers (Marburg, Ebola, Lassa, Lujo, etc.) so blood samples were sent away to be tested. Tests for a range of human pathogens that would cause a similar illness proved negative.

A pathogen discovery protocol was then put in place to see if her disease was caused by a new infection. Techniques of non-specific deep-sequencing and computer-based sequence analysis like those used to discover the novel cyclovirus described above revealed that she was indeed infected with a novel paramyxovirus. This virus had never been seen before and was most closely related to viruses previously isolated from fruit bats in China and Ghana (Figure 4.3). It proved possible to isolate the virus by infecting mice then propagating viruses present in mouse brains.

This story exemplifies several important points in emerging infections. First, the huge power of modern genetic sequencing technologies to detect novel human pathogens. Second, the importance of 'sentinel populations' – people who have extraordinary levels of exposure to potential human pathogens – as a source of early warnings about transmission from animals to humans of infections that we have not yet discovered. Third, that the species barrier is not the defining threshold for emerging infections. Single cases that cause no secondary infections are medically important, but pose no threat to public health. The defining threshold is the ability of an emerging infection to cause sustained human-to-human transmission.

February 2014: Human Infections with Avian Influenza Viruses in China

In February, the European Centres for Disease Control (ECDC) published a risk assessment concerning human infection with bird-derived influenza

FIGURE 4.3 Many emerging infections are transmitted to people from wild animals. Examples of this process described here include: (a) *Cryptosporidium* from hedgehogs (*Erinaceus europaeus*). © Michael Gäbler/Wikimedia Commons/CC-BY-SA-3.0 (b) MERS CoV from Dromedary Camels (*Camelus dromedaries*)© Peretz Partensky/ Wikimedia Commons (c) Susoga virus perhaps from fruit bats (*Eidolon helvum*) reservoir of the most closely related virus of African origin. © Fritz Geller-Gromm/Wikimedia Commons.

viruses in China.[28] Ever since 1997, when six people died of H5N1 avian influenza in Hong Kong, there has been rumbling concern about the potential for a bird-derived influenza to cause a pandemic. This concern focused upon H5N1 influenza for many years. From 2003 until May 2014, 665 cases of H5N1 from 16 countries around the world were confirmed, 392 of whom died. In China, over the same period, there were 46 cases of H5N1 influenza, 30 of whom died.

In the spring of 2013, China reported cases and deaths caused by a different influenza, designated H7N9. By February 2014 there had been 354 H7N9 influenza cases in China, including 113 deaths. This is 100-fold more H7N9 cases than H5N1 cases in China over an equivalent time period.

For the two years that H7N9 has been observed there seems to be a strong seasonality, with all but a handful of cases occurring during the winter. All known cases have been acquired in China and the great majority of infected people have had some contact with poultry or live-bird markets.

Increased surveillance for avian influenza in China led to the detection of cases with two further types of influenza: three cases infected with H10N8 influenza and one with H6N1. Neither of these types of influenza had ever been seen to infect humans before.

H5N1 and H7N9 influenza share the following characteristics. In humans the great majority of cases are acquired from poultry. There is evidence for occasional human-to-human transmission, but not for sustained chains of transmission. Both infections have a high case fatality rate in humans: 59 per cent for H5N1 and 32 per cent for H7N9.

However, there are also differences. H7N9 infections are currently confined to China, (with just one exported case) whilst human H5N1 infections have occurred in sixteen different countries. H7N9 cases show a strong pattern of seasonality, whilst H5N1 infections in China do not. The two infections seem to have different age distribution (mean age at infection fifty-five years for H7N9 and twenty-eight years for H5N1), and there are more H7N9 infections in men, with no such bias for H5N1. However, the major difference is the much greater number of H7N9 cases: since 2003 there have been 46 cases of H5N1 in China and since 2013 there have been 354 cases of H7N9.

Why is there suddenly so much avian influenza in China? The small numbers of rare infections (H10N8 and H6N1) can most likely be attributed to heightened surveillance. The large number of H7N9 cases is much more worrying. H7N9 infection does not cause severe disease in poultry (unlike H5N1). This has the unfortunate effect that the first sign of infection of a poultry flock can be cases of severe disease in humans. One explanation for the growing number of human cases in China is, therefore, that there is a hidden zoonotic epidemic with sporadic transmissions to humans. Under this scenario the greatest threat from H7N9 is further spread amongst poultry – possibly to other countries.

There is also the lingering concern that H7N9 influenza might, through mutation, acquire the ability to cause long chains of human-to-human transmission.[10,11,29] To our knowledge no H7-type influenza virus has ever circulated widely in humans, so the pool of susceptibles would be very large. A large susceptible pool, a virus with efficient human-to-human transmission and a high case fatality rate would be a recipe for disaster (Figure 4.3).

March 2014: Ebola Virus Disease in West Africa

On the 23 March the Ministry of Health in Guinea (West Africa) notified WHO of a 'rapidly evolving outbreak of Ebola virus disease (EVD)'. At that time forty-nine cases including twenty-nine deaths had been reported. That was already more cases than the median-sized outbreak (forty-four) in the twenty-four outbreaks since the first description of EVD in 1976 (Figure 4.4).

EVD is a severe viral infection with an unusually high case fatality rate. The twenty-nine deaths from forty-nine cases by March 2014 yield a case fatality rate of 60 per cent and this is not abnormal; of the ~2400 cases that occurred between 1976 and 2012 (see Figure 4.4) nearly 1600 died.

EVD is not endemic in humans. Outbreaks arise when people become infected after contact with the wild animals that are the natural reservoir. Illness is characteristic of a viral haemorrhagic fever: fever, fatigue and signs of both internal and external bleeding. Transmission from human to human happens through close contact with infected patients or their bodily secretions. There are serious problems with transmission of EVD in hospitals or during funeral ceremonies which include close contact between mourners and the body of the deceased. At the beginning of this outbreak in 2014 there was no vaccine and no specific anti-viral treatment.

The EVD outbreak in West Africa continued to evolve through the spring of 2014 and spread to Sierra Leone and Liberia. In late May 2014 there was a marked rise in the number of cases. By late June 2015 there had been over 27,000 cases and over 11,000 deaths,[31] making this the largest ever outbreak of EVD. The fear, dimensions and global concerns that this epidemic raised and how it changed medical processes are discussed in Chapter 1.

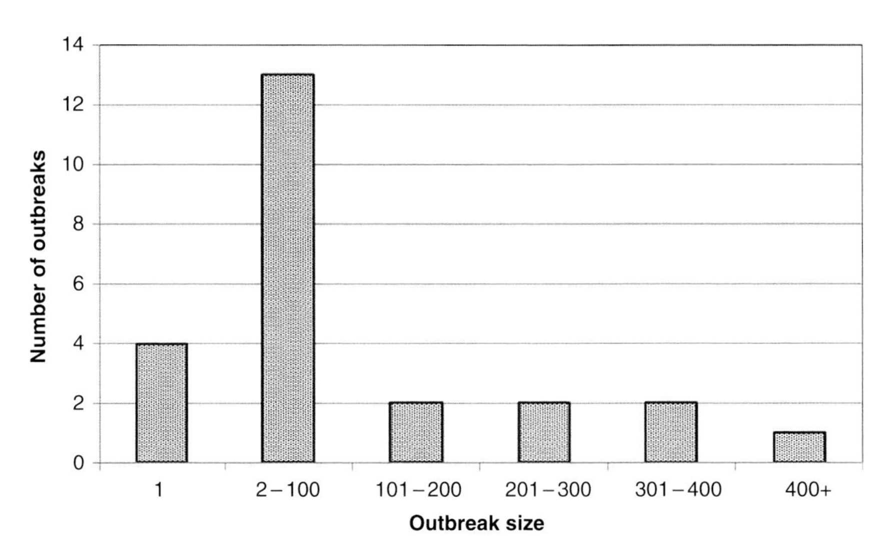

FIGURE 4.4 Outbreak size distribution for twenty-four outbreaks of Ebola virus disease between first description in 1976 and 2012.[30]

April 2014: Antimicrobial Resistance

April's story tells not of a single disease outbreak but of the slowly unfolding global disaster of the evolution of antimicrobial resistance. The extent of the spread of drug-resistant pathogens is dramatically highlighted by WHO's first global report on surveillance for antimicrobial resistance. The report[2] (WHO 2014a), published in April 2014, focuses upon seven common bacterial infections, and finds evidence of widespread antibacterial resistance for all of them.

For example, *Staphylococcus aureus* is a bacterium that can cause a wide range of skin, bloodstream and bone infections. When resistance to penicillin evolved in the 1940s new drugs were developed that overcame this resistance. *Staphylococcus aureus* strains resistant to these new drugs (called methicillin-resistant *Staphylococcus aureus* or MRSA) first emerged in the 1960s and have now spread around the world. The newly published survey found that in five of six WHO regions at least one country reported national data in which more than half of *Staphylococcus aureus* isolates are methicillin resistant. Patients with such infections are more difficult and more expensive to treat. This pattern of widespread resistance is the norm for bacteria that are commonly acquired in a hospital setting. It does not mean that half of all *Staphylococcus aureus* is methicillin resistant, but it does mean that in most of the world there is at least one country where a national survey reported such high levels of resistance.

A global survey of this nature must, perforce, cope with varying levels of data quality. For example, South East Asia is the only WHO region with no country reporting >50 per cent MRSA in national data, but only three countries from that region provided national data. There are other sources of variability, like the size of the sample and the kind of patient from whom isolates were collected. Indeed, the second key finding of the whole study is that there are significant gaps in the data.

However, the first key finding is that there are already very high rates of drug resistance in common bacterial infections across the globe. This leads to the following gloomy prognostication in the report's foreword:

> A post-antibiotic era – in which common infections and minor injuries can kill – far from being an apocalyptic fantasy is a very real possibility for the 21st century.

May 2014: Polio, a Public Health Emergency of International Concern

The most dramatic emerging infectious disease event of the twelve months described here was the May 2014 declaration of a 'Public Health Emergency of International Concern' (PHEIC) over the international spread of Polio. This is only the second time a PHEIC has been issued since they were introduced under the International Health Regulation in 2005. The first PHEIC was issued during the early circulation of H1N1 influenza in what eventually became the 2009 pandemic.

The 'emergency' label was activated because during the first four months of 2014 Polio had spread out of three different countries: from Pakistan to Afghanistan, from Syria to Iraq and from Cameroon to Equatorial Guinea. January to April would normally be the low transmission season, so there was particular concern that the onset of the high transmission season in May and June might lead to further international spread.

Data at 18 June 2014 showed ongoing problems in Pakistan, with 82 cases in the year-to-date (of a global case-count of 103) of which 7 cases had been reported in the preceding week. The other two exporting countries had, at 18 June, had no new cases since January.[32]

There is a horrible irony in the fact that the one infectious disease event that was deemed an 'international emergency' of these twelve months was caused by a once-common childhood infection against which there is a cheap, safe and (largely) effective vaccine. The re-emergence of polio is a lesson to us all that keeping such infections under control requires constant vigilance.

Discussion

Across the twelve months recorded here emerging infectious disease events of all types occurred. Four infections new to man were described: the novel virus attacking the central nervous system of patients in Vietnam; the hedgehog-associated *cryptosporidium* from an individual in the Czech Republic; Sosuga the new virus from a wildlife biologist and

MERS-CoV from the ongoing outbreak in the middle East. There were three examples of infections detected in new geographic regions: EVD in West Africa; Chikungunya in the Caribbean and Zika virus in French Polynesia. Two infectious diseases re-emerged: Cholera in Mexico and Polio in multiple countries with unusual examples of cross-border infection. There were two examples of emerging new variants of known infectious agents: H7N9 influenza in China and the global spread of antimicrobial pathogens of many sorts. Finally, there was just one case of a new association between a known pathogen and a known disease: the link between Hepatitis E virus infection and Guillain-Barré syndrome.

If, instead of picking one event from each month, one were to review all events of the past twelve months the impression would be one of constant activity. New infections are discovered, old infections spread to new places and diseases that were once controlled re-emerge as control efforts fail. But are there more emerging infectious diseases than there were in the past? In terms of cases of infection with agents that would be classed as 'emerging' there probably are more than there were in the past. There are three drivers for this. First, there are more people, more of whom live in densely populated cities. Second, those people mix more freely, allowing infections that would once have been confined to small villages to spread widely. Third, our collective ability to detect and characterise infectious agents has taken a step change upwards in recent years. The metagenomic techniques that allow the identification of pathogens that cannot be cultured allows the identification of infectious agents that would simply have been invisible in years gone by.

Although there are more emerging infectious disease events, it is not obvious that there are more infections. For deaths from infection the trend is clearly in the opposite directions. Figure 4.5 records adult male mortality by cause in Great Britain through the twentieth century. Deaths from infections (in orange) pick up at the beginning of the First World War in 1914, peak with the Spanish Influenza pandemic of 1918 (with an even larger peak among deaths from respiratory causes in red), pick up again at the start of the Second World War in 1939, then melt away with the introduction of antibiotics in the mid-1940s.

Around the world similar patterns are coming into play.[33] In 1990, communicable diseases accounted for 25 per cent of global deaths; by 2010 this had fallen to 19 per cent. However, this global figure underemphasises the continuing importance of infections as a cause of death for many poorer parts of the world; in low-income countries, communicable diseases are still responsible for one-third of all deaths.

For three of the events of the past twelve months the ready availability of a pool of susceptible individuals played a large role in generating an outbreak. Both Zika virus and Chikungunya are insect-transmitted diseases that have recently moved into new geographic areas. Because they have not circulated in these locations the population is not immune. Since the transmitting insect is present this has, in both cases, lead to explosive outbreaks. Polio outbreaks too are caused by the presence of too many susceptible individuals. However, in the case of polio it is failure to vaccinate young children that has caused the pool of susceptible individuals to grow large enough to trigger outbreaks.

Thus the distribution of people around the world and their immune status with respect to different infections is crucial information for understanding where risks for the spread of new infections will lie.

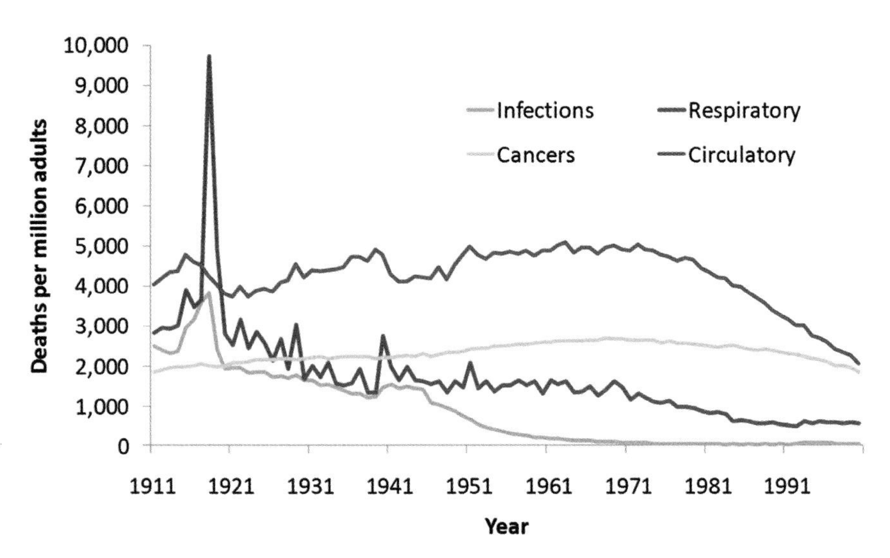

FIGURE 4.5 Adult male mortality by cause, Great Britain 1911–2000. Figures are deaths per million adults aged 15–74. From orange – infectious disease, red – respiratory disease, yellow – cancers, blue – circulatory disease.
(Data Source: Office for National Statistics)

Figure 4.6 sketches population data per square kilometre around the world. As many people live inside the circle centred upon South East Asia as live outside it. Some authors have identified this part of the world as a 'hot spot' for generating emerging infections.[34] But since so many humans live there, it is, perhaps, not surprising that many infections of humans should arise there.

The big fear for emerging infectious diseases is of a global pandemic caused by a novel infectious agent that transmits well, spreads fast, has a high case fatality rate and for which there is neither vaccine nor cure. It is for fear of such an event that the progress of H7N9 influenza and MERS-CoV in the Middle East are so carefully monitored, watching in case either infection were to gain the ability to transmit well from human to human. However, some argue[36] that the evolution of antibiotic resistance is the more dangerous threat, and a relatively indefensible one that is already amongst us.

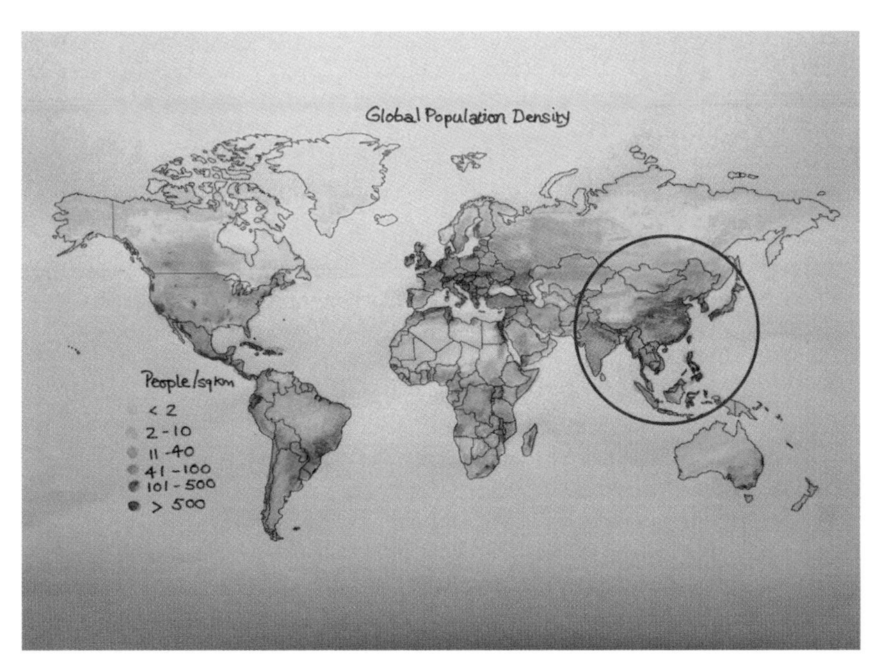

FIGURE 4.6 Global population density. As many people live inside the brown circle as live outside it. (redrawn from Center for International Earth Science Information Network maps, available online[35]).

References

1. Public Health England (2014) Emerging Infections Monthly Summaries. www.hpa.org.uk/webw/HPAweb&Page& HPAwebAutoListName/Page/1234254470752 (accessed June 2014).
2. WHO (2014a) Antimicrobial Resistance. Global Report on Surveillance. www.who.int/drugresistance/documents/surveillancereport/en/, http://apps.who.int/iris/bitstream/10665/112642/1/9789241564748_eng.pdf?ua=1 (accessed 18 November 2016).
3. Le Van Tan H., Doorn H.D.T.N., Tran Thi Hong Chau L.T., Phuong Tu M.D. V., Marta Canuti M.D., Maarten F.J., Menno D. (2013) Identification of a new cyclovirus in cerebrospinal fluid of patients with acute central nervous system infections. *mBio* 4(3):e00231-13.
4. ECDC (2013a) Rapid Risk Assessment. Novel Cyclovirius CyC-VN. www.ecdc.europa.eu/en/publications/Publications/rapid-risk-assessment-Cyclovirus-final.pdf (accessed 18 November 2016).
5. de Jong M.D., Van Kinh N., Trung N.V., Taylor W., Wertheim H.F., van der Ende A., van Doorn H.R. (2014) Limited geographic distribution of the novel cyclovirus CyCV-VN. Scientific reports, 4 3967.
6. Chen K., Pachter L. (2005) Bioinformatics for whole-genome shotgun sequencing of microbial communities. *PLoS Comput Bio.* 1(2):e24. doi:10.1371/journal.pcbi.0010024.
7. WHO (2013) WHO Statement on the Second Meeting of the IHR Emergency Committee concerning MERS-CoV. www.who.int/mediacentre/news/statements/2013/mers_cov_20130717/en/ (accessed 10 June 2014).
8. ECDC (2014a) Rapid Risk Assessment Zika virus outbreak, French Polynesia. http://ecdc.europa.eu/en/publications/Publications/Zika-virus-French-Polynesia-rapid-risk-assessment.pdf (accessed 18 November 2016).
9. WHO (2014b) WHO statement on the Fifth Meeting of the IHR Emergency Committee concerning MERS-CoV. www.who.int/mediacentre/news/statements/2014/mers-20140514/en/ (accessed 10 June 2014).
10. Arinaminpathy N., McLean A.R. (2009) Evolution and emergence of novel human infections. *Proc. R. Soc. B.* 273:3075–83.
11. Kubiak R.J., Arinaminpathy N., McLean A.R. (2010) Insights into the evolution and emergence of a novel infectious disease. *PLoS Comput Biol.* 6(9):e1000947
12. Santos L., Mesquita J.R., Rocha Pereira N., Lima-Alves C., Serrão R., Figueiredo P., Reis J., Simões J., Nascimento M.S., Sarmento A. (2013) Acute hepatitis E complicated by Guillain-Barré syndrome in Portugal, December 2012 – a case report. *Euro Surveill.* 18(34):pii=20563.

13. Kamar N., Bendall R.P., Peron J.M., Cintas P., Prudhomme L., Mansuy J.M. et al. (2011) Hepatitis E virus and neurologic disorders. *Emerg Infect Dis.* 17(2):173–9. http://dx.doi.org/10.3201/eid1702.100856. PMid:21291585. PMCid:PMC3298379.
14. Scharn N., Ganzenmueller T., Wenzel J.J., Dengler R., Heim A., Wegner F. (2013) Guillain-Barré syndrome associated with autochthonous infection by hepatitis E virus subgenotype 3c. *Infection* 42(1): 171–3.
15. WHO Global Health Observatory Data Repository. http://apps.who.int/gho/data/node.main.175 (accessed 18 November 2016).
16. Bartlett S., (2014) Infectious disease surveillance update. *Lancet Infect Dis.* 14:19, ISSN 1473–3099, http://dx.doi.org/10.1016/S1473-3099(13)70372-4.
17. Moore S.M., Shannon K.L., Zelaya C.E., Azman A.S., Lessler J. (2014) Epidemic risk from cholera introductions into Mexico. *PLOS Curr Outbreaks.* Edition 1. doi: 10.1371/currents.outbreaks.c04478c7fbd9854ef6ba923cc81eb799.
18. Kvac, M. et al. (2013) Gastroenteritis caused by the Cryptosporidium hedgehog genotype in an immunocompetent man. doi: 10.1128/JCM.02456–13 JCM.02456–13.
19. Duffy M.R., Chen T.H., Hancock W.T., Powers A.M., Kool J.L., Lanciotti R.S, et al. (2009) Zika virus outbreak on Yap Island, Federated States of Micronesia. *N Engl J Med* 360(24):2536–43.
20. Direction de la Santé BdVs, Polynesie Francaise.(2014) Surveillance de la dengue et du zika en Polynésie Française. www.hygiene-publique.gov.pf/spip.php?article120 (accessed 18 November 2016).
21. Pialoux G., Gauzere B.A., Jaureguiberry S., Strobel M. (2007) Chikungunya: an epidemic arbovirosis. *Lancet Infect Dis.* 7(5):319–27.
22. Vazeille M., Moutailler S., Coudrier D., Rousseaux C., Khun H., Huerre M. et al. (2007) Two Chikungunya isolates from the outbreak of La Reunion (Indian Ocean) exhibit different patterns of infection in the mosquito, Aedes albopictus. *PLoS One.* 2(11): e1168.
23. Rezza G., Nicoletti L., Angelini R., Romi R., Finarelli A.C., Panning M. et al. (2007) Infection with chikungunya virus in Italy: An outbreak in a temperate region. *Lancet* 370(9602):1840–6.
24. Grandadam M., Caro V., Plumet S., Thiberge J.M., Souares Y., Failloux A.B. et al. (2011) Chikungunya virus, southeastern France. *Emerg Infect Dis.* 17(5):910–13.
25. ECDC (2013b) Rapid Risk Assessment Chikungunya fever, Saint Martin. www.ecdc.europa.eu/en/publications/Publications/

chikungunya-st-martin-rapid-risk-assessment.pdf (accessed 18 November 2016).
26. PAHO (2014) Number of Reported Cases of Chikungunya Fever in the Americas. www.paho.org/hq/index.php?option=com_content&view=article&id=9053&Itemid=39843 (accessed 18 November 2016).
27. Albariño C.G., Foltzer M., Towner J.S., Rowe L.A., Campbell S., Jaramillo C.M. et al. (2014) Novel paramyxovirus associated with severe acute febrile disease, South Sudan and Uganda, 2012. *Emerg Infect Dis.* DOI: 10.3201/eid2002.131620.
28. ECDC (2014b) Rapid Risk Assessment Human infection with avian influenza viruses in China. www.ecdc.europa.eu/en/publications/Publications/avian-flu-china-rapid-risk-assessment-26022014.pdf (accessed 18 November 2016).
29. Nicoll A., Danielsson N. (2013) A novel reassortant avian influenza A (H7N9) virus in China – what are the implications for Europe. *Euro Surveill.* 18(15):20452.
30. WHO (2014c) Ebola Virus Disease in Guinea. www.who.int/csr/don/2014_03_23_ebola/en/. (WHO 2014d) Ebola Virus Disease. www.who.int/mediacentre/factsheets/fs103/en/(accessed 18 November 2016).
31. WHO (2015) Ebola Situation Reports. http://apps.who.int/ebola/ebola-situation-reports (accessed 18 November 2016).
32. Global Polio Eradication Initiative (2014) Polio this week. www.polioeradication.org/Dataandmonitoring/Poliothisweek.aspx (accessed 18 November 2016).
33. Lozano, R. et al. (2013) Global and regional mortality from 235 causes of death for 20 age groups in 1990 and 2010: A systematic analysis for the Global Burden of Disease Study 2010. *Lancet* 380:2095–2128, ISSN 0140–6736, http://dx.doi.org/10.1016/S0140-6736(12)61728-0.
34. Jones KE, Patel NG, Levy MA, Storeygard A, Balk D, Gittleman JL, Daszak P. (2008) Global trends in emerging infectious diseases. *Nature* 21;451(7181):990–3.
35. http://sedac.ciesin.columbia.edu/data/collection/gpw-v3/maps/gallery/browse (accessed 18 November 2016).
36. DoH (2013) Annual Report of the Chief Medical Officer Volume 2. www.gov.uk/government/publications/chief-medical-officer-annual-report-volume-2 (accessed 18 November 2016).

Further Reading

Assiri A1, Al-Tawfiq JA, Al-Rabeeah AA, Al-Rabiah FA, Al-Hajjar S, Al-Barrak A, Flemban H, Al-Nassir WN, Balkhy HH, Al-Hakeem RF, Makhdoom HQ, Zumla AI, Memish ZA. (2013). Epidemiological, demographic, and clinical characteristics of 47 cases of Middle East respiratory syndrome coronavirus disease from Saudi Arabia: a descriptive study. Lancet Infect Dis;13(9):752–61.

5 Plagues, Populations and Survival

STEPHEN J. O'BRIEN

The rush of whole genome sequences spurred by the human genome project has heralded a new way to explore unknown events in our pre-history. Using gene sequences from the genome we are able to mine the vast amount of genetic information coming out of genome sequences from humans and animal species. Certain genetic patterns or 'footprints' allow us to deduce ancient defining events in the natural history of a species. Just as early palaeontologists dug up the fossil remains of extinct species, 'genetic archaeologists' are beginning to reconstruct the origins of genomic patterns and to link them to ancient demographic events. Modern diseases that clog wards in the world's hospitals represent a major challenge to human health. Historically, such diseases were a selective regulator on a species, the genetically strong individuals survived an infectious disease outbreak and the weak perished. Those that survived passed on the good, beneficial genes to their descendants who later had the right genetics that gave their immune systems an advantage the next time the same type of disease came along. Past interactions between pathogens which had genes that allowed them to quickly adapt finely tuned ever-evolving immune defences that we and other species have developed to protect us from pathogenic organisms. This represents a biological puzzlement that is only beginning to be deciphered. Here, I illustrate how we learn of the acquired, hidden lessons of survival, adaptation and genome evolution by examining natural history from within the genomes of animals and humans. In so doing, we are able to provide an early glimpse of the coming discipline of genomic archaeology. I will illustrate this with examples from 'The Plague' (*Yesinia pestis*, the cause of the black death), SARS and HIV/AIDS. Furthermore, I will describe how scientists have been able to track the emergence and progression of deadly outbreaks, sometimes revealing unfathomed threats to the very existence of a population or the whole species. Examples will include how this new science is able to unlock medical secrets that someday may be used to prevent the extinction of a species.

Infectious diseases, genomes and the arms race

The delicate balance between host species and their pathogens is akin to a deadly arms race, waged fiercely every day multiplied by individuals and populations in different geographical settings. These events are to a different extent recapitulated over the approximately 62,000 species of vertebrates that survive on earth today. The struggle is frequently lost, never won, but temporarily circled with the survivors rising once again and compete another day. I can still recall back in 1967 when US Surgeon General William Stewart proudly pronounced an end to the horrors of infectious diseases. Encouraged by success of antibiotics and vaccines against smallpox, measles, mumps and polio he predicted a swift shift in biomedical emphasis from acute infections to chronic diseases. He was so wrong. Soon came hepatitis B, hepatitis C, HIV-1 and 2, Hantavirus, Influenza, Ebola, SARS, West Nile and Mad Cow diseases[1] to name but a few. Modern medicine has far from conquered infectious diseases, and with the new threat of anti-microbial and vaccine resistance and re-emerging viral diseases, the war is far from over.

I cannot begin to review all the provocative new insights so recently revealed by the technological revolution brought to us in the 'Genomics Era'. Human scourges have extracted an enormous loss of life throughout human history; from smallpox to leprosy to bubonic plague. Modern medicine continues to develop better diagnostics, preventive vaccines and therapeutics. However, we are a long way from success. In effect we are but a short step ahead of the next epidemic (Table 5.1). Twentieth-century outbreaks are terrifying when we realise how few we have successfully detected early enough to prevent, and how few medicines or vaccines we have ready to defend against the next emerging disease threat.

The genomics era has given us a new perspective, one which involves examining genetic diversity of the host species when a new pathogen appears. A few decades ago we knew of only one gene variant that protected against an infectious disease: one copy of the sickle cell anaemia gene variant HBB^S protects its carriers from malaria while two copies of the same haemoglobin gene mutation causes

Table 5.1 *World Pandemics of the Twentieth Century.*[1]

• 1800–2000	Tuberculosis
• 1918	Spanish Flu
• 1950	Polio
• 1982	HIV-AIDS
• 2003	SARS

sickle cell anaemia. Today the number of gene variants that have been identified that influence the outcome of different diseases is in the hundreds. To illustrate how the genes of an individual and population determine the response to diseases I recount three stories from my personal experience. Each narrative provides a special lesson for recognition and public health defences against contagions of the future. My first example goes back to November, 2002 in Guangdong Province in southern China.

Severe Acute Respiratory Syndrome – SARS

The severe acute respiratory syndrome (SARS) outbreak first appeared as a flu-like disease in hospitals of Hong Kong and Guangdong Province in Southern China in late 2002.[2,3,4] Within nine months the virus spread to 29 countries, infected over 8000 people and caused nearly 800 deaths[5]. The cause was determined to be a new coronavirus (akin to some viruses which can cause the common cold in humans) that spread with alarming speed among health care workers, through casual contacts, and across the globe. The result was human suffering, large scale quarantine, restricted human movement and huge economic costs. Interestingly, virus sequences that were closely related to SARS virus sequences from patients were obtained from Himalayan palm civets, *Paguma lavarta*, collected in Chinese food markets.[6] Further screening of wildlife more recently identified horseshoe bats as a natural carrier species for SARS coronaviruses (SL-CoV).[7,8,9] The epidemic was contained by May 2003 as the result of extremely strict quarantine measures. While we now know much about the virus, and while there are promising developments in the search for anti-viral drugs to combat the SARS virus,[10,11] there are still no vaccines or effective treatments for infected patients.

The intensity of the SARS outbreak caught the global public health community by surprise. This new virus had unexpectedly emerged with deadly consequences. Veterinary virologists were more familiar with coronaviruses in livestock, dogs, cats and poultry but for the most part found these viruses seldom cause fatal diseases in domestic animals.[4,12] However, exceptions have occurred; for example in pigs a single nucleotide variant of porcine coronavirus was known to give rise to virulent pathogenic enteric coronavirus infections.[13,14] A second exception involved a devastating feline coronavirus outbreak in cheetahs (*Acinonyx jubatus*) documented in a wild animal park in rural Winston, Oregon, four decades ago.[15,16]

At that time the Wildlife Safari organisation was the most prolific cheetah breeding facility in the world, holding some sixty cheetahs. In May 1982, two young cheetahs arrived from the Sacramento Zoo in California with symptoms of fever, severe diarrhoea, jaundice and neurological spasms. Both died and were diagnosed by veterinary pathologists with a common form of feline infectious peritonitis, a disease in domestic cats caused by coronavirus, Feline infectious peritonitis virus (FIPV).[12,15] Within six months, every cheetah in the park became infected and developed antibodies to FIPV. By 1984, 60 per cent of the cheetahs had died of infectious peritonitis.[15,16,17] To our knowledge, this was the worst outbreak with the highest mortality of infectious peritonitis in any cat species; in domestic cats mortality caused by this virus seldom exceeds 5 per cent.

At the time of the Oregon infectious peritonitis outbreak there were no molecular diagnostic tools, such as the PCR assay, which are commonly used in clinical diagnostic microbiology labs today. When SARS appeared in 2003, archival frozen specimens were thawed to re-examine the origins of the cheetah coronavirus.[18] DNA sequences of three viral genes from five cheetahs that had died of infectious peritonitis revealed that the viruses that killed the cheetahs were close relatives of domestic cat FIPV, as well as other well-known coronaviruses such as porcine transmissible gastroenteritis virus (TGEV) and canine coronavirus (CcoV) (Figure 5.1).

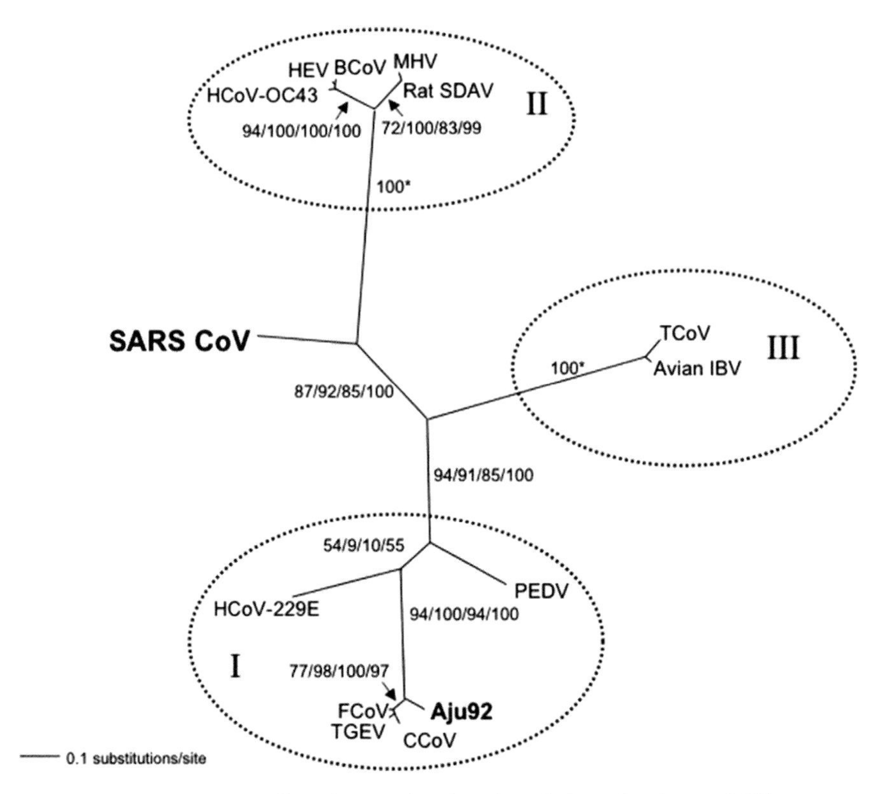

FIGURE 5.1 Genetic map showing the relative relatedness of different coronaviruses including the cheetah coronavirus (CCoV, Aju92). This was deduced from sequences of archived liver and kidney tissues from cheetahs living at Wildlife Safari during the coronavirus outbreak.[18] Numbers plotted along the branches indicate bootstrap values and Bayesian posterior probabilities shown as percentages depicted in the following order, Maximum Likelihood/Maximum Parsimony/Minimum Evolution/Bayesian. The three major coronavirus antigenic groups are supported by the genomic data and are indicated by hatched circles and roman numerals. Abbreviations are as follows: human coronavirus 229E (HCoV-229E), canine coronavirus (CCoV), feline coronavirus (FCoV), CCoV Aju92, porcine transmissible gastroenteritis virus (TGEV), porcine epidemic diarrhoea virus (PEDV), human coronavirus OC43 (HCoV-OC43), bovine coronavirus (BCoV), porcine hemagglutinating encephalomyelitis virus (HEV), rat sialodacryoadenitis (SDAV), mouse hepatitis virus (MHV), turkey coronavirus (TCoV), avian infectious bronchitis virus (avian IBV), SARS coronavirus from human (SARSCoV) and from palm civet (SZ16).

Furthermore, when the genetic sequences were compared, the cheetah strains were quite closely related to FIPV strains isolated from domestic cats with very few differences between them. They were almost identical to the coronavirus that killed cheetahs and the more innocuous domestic cat FIPV. Given the high genetic similarity between domestic cat FIPV and cheetah FIPV, plus the fact that several lions (*Panthera leo*) at Winston Park became infected but did not succumb to disease, we suggested the reason for the extremely high morbidity and mortality was due to the genetics of cheetahs and that they were immunologically naïve, likely not having been exposed to this domestic cat virus before.

Cheetahs are well known as one of the most genetically homogeneous species amongst mammals.[16] They descended from a catastrophic extinction event that eliminated scores of large mammal species in North America, Europe, Asia and Australia 10,000–12,000 years ago.[19] The fortunate survivors were subsequently subjected to generations of close inbreeding during the cheetah's ancestry, reducing overall genomic diversity 10–100 fold. This genetic impoverishment would reduce variation in the immune response genes within the major histocompatibility complex-MHC of the genome.[16,19,20,21] The mammalian MHC region of the genome encodes over sixty tightly linked extremely variable genes, offering a diverse adaptive repertoire for immune recognition of invading pathogens. The basis for MHC-linked disease resistance is its genetic diversity within a population. Simply put, it provides a broad net to capture the moving target of pathogens, for their recognition and removal. Thus, when a microbe evolves a strategy to escape immune defences of one individual, others in a genetically diverse population may still be protected and survive. But, for the Oregon cheetahs, the conditions for widespread disease in the population were set with infection of the first individual since others within this population were vulnerable by shared genetic and immunologic homogeneity; they were all at risk. It was as if they were all immunological clones. Stated illustratively, since this virus had escaped the immunological net of the first individual, and the others were near identical, there was very little to stop the virus in this population.

The cheetah outbreak taught us some important lessons for wildlife and for SARS. First, the human SARS coronavirus (SARS-CoV) was a novel strain that showed a long genetic distance from the known human coronavirus isolates (Figure 5.1). The SARS-CoV has likely evolved in bat species undetected for generations. By contrast the cheetah coronavirus was genetically indistinguishable from the ordinary Feline infectious peritonitis (FIP) coronavirus we know in domestic cats. Second, as bad as SARS became the overall mortality was relatively low (<10 per cent), as is also observed in FIP domestic cat outbreaks (<5 per cent mortality[12]). Infected cheetahs, however, exhibited the opposite extreme with 90 per cent morbidity and over 60 per cent mortality. The combined evidence suggests that host genetic uniformity makes epidemics (epizootics in the case of animals) much worse. It is likely that variation in immune defence genes of genetically outbred humans and domestic cats has protected them from a deadly disease, while most of the genetically similar cheetahs have succumbed to disease and death in the majority of animals. We know from textbooks that inbreeding in livestock and inbred rodent strains gives increased numbers of disease-susceptible animals following a viral outbreak. However, it took the lesson from the cheetah to remind us how important abundant genetic diversity in immune defence really becomes when a pathogenic virus threatens a natural population.

Feline Immunodeficiency Virus: FIV – A Natural AIDS Model

AIDS had first appeared in the early 1980s as a clustering of patients from homosexual communities inflicted with a rare cancer, Kaposi's sarcoma, and *Pneumocystis carnii* pneumonia.[22] The contagion quickly began to spread among recipients of blood transfusions, particularly among haemophiliacs. AIDS is caused by a lentivirus (see Figure 5.2), a genus within the family Retroviridae, termed human immunodeficiency virus (HIV) that infects and destroys CD4-bearing T-cells. There are two distinct HIV viruses (types 1 and 2) and since its recognition in the early 1980s, HIV-1 has spread around the globe infecting over 78 million people and killing over 35 million (www.unaids.org). Soon after its discovery, AIDS became the deadliest disease in recorded history; it was a disease with no vaccine or treatment until the mid-1990s. While there have been tremendous advances in anti-viral therapy, treatment

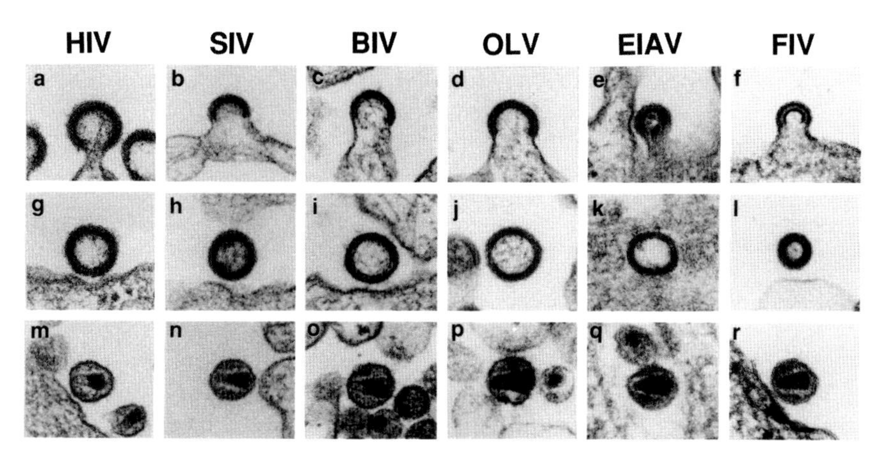

FIGURE 5.2 Electron microscope images of different lentiviruses from human, monkey, cow, sheep, horse and cat, Micrographs courtesy of DM Gonda. http://ncifrederick.cancer.gov/atp/imaging-and-nanotechnology/electron-microscopy-laboratory/eml-protocols-and-resources/eml-image-gallery/retrovirus3/

simply inhibits virus replication but does not clear the DNA copies of the virus from the infected patients. We are far from finished with HIV/AIDS, which infected between 2.1–2.7 million new victims per year between 2012 and 2016 (www.unaids.org).

Lentiviruses are well known to affect many animal species such as sheep goats, horses, cattle and cats (feline immunodeficiency virus, FIV), (Figure 5.2). With the exception of some primate lentiviruses in certain species, only FIV in domestic cats causes AIDS-like disease. FIV was first discovered in 1986 in a house cat with a wasting-like disease.[23] FIV has a small viral RNA genome with many similarities with the human version HIV-1. FIV is epidemic among feral domestic cats and other Felidae species throughout the world, having diverged into several phylogenetic clade types across continents.[24,25] FIV infection leads to an AIDS-like syndrome of immune depletion, increased susceptibility to rare cancers and common infections causing death (Table 5.2).

Once it became clear that FIV caused AIDS in feral domestic cats, we wondered if the virus had spread to other species of Felidae. As shown in

Table 5.2 *Symptoms of FIV-AIDS in the domestic cat.*[60]

FIV antibody positive
Early flu-like symptoms
Severe weight loss-wasting disease
Depletion of CD4 T-lymphocytes
Respiratory infections
Skin lesions
Fungal/bacterial infections
Rare cancers: lymphomas
Neurological tremors and neuropath

Figure 5.3, a comprehensive survey of archived serum samples revealed that FIV had infected many of the thirty-seven described species of the felidae family (Figure 5.3).[25]

Genetic analysis of individual FIV-variants in a dozen or more species of felids[15,24,30,31] revealed a relationship of FIV among various species (Figure 5.4) (that is, every lion strain has as its closest relative another lion isolate rather than FIV from a different cat species (Figure 5.4)).

These results supported the notion that, although FIV occasionally can move from species to species, these events were exceedingly rare, leading to an expansion of viral genome sequence diversity within the different cat species, so that most carried their own distinct version of FIV. Interestingly, our genetic analysis revealed that most cat species became infected by FIV variants within the last 10,000–20,000 years. Thus, much older than when HIV-1 or HIV-2 entered humans to cause AIDS. Interestingly, patterns of evolution within both virus and host genomes suggest FIV may have existed far longer in some species.[26]

At first glance, there seemed to be no clear clinical pathology among FIV-infected felids in zoological collections or in seemingly healthy (or asymptomatic) FIV-infected cats in natural populations of felids. This suggested that FIV was pathogenic in domestic cats but non-pathogenic or innocuous in other wild Felidae species.[25] That conclusion was tentative for many species, but certainly did not hold up upon closer scrutiny of FIV-infected lions.[27,28,29] Our team mounted a broad and comprehensive

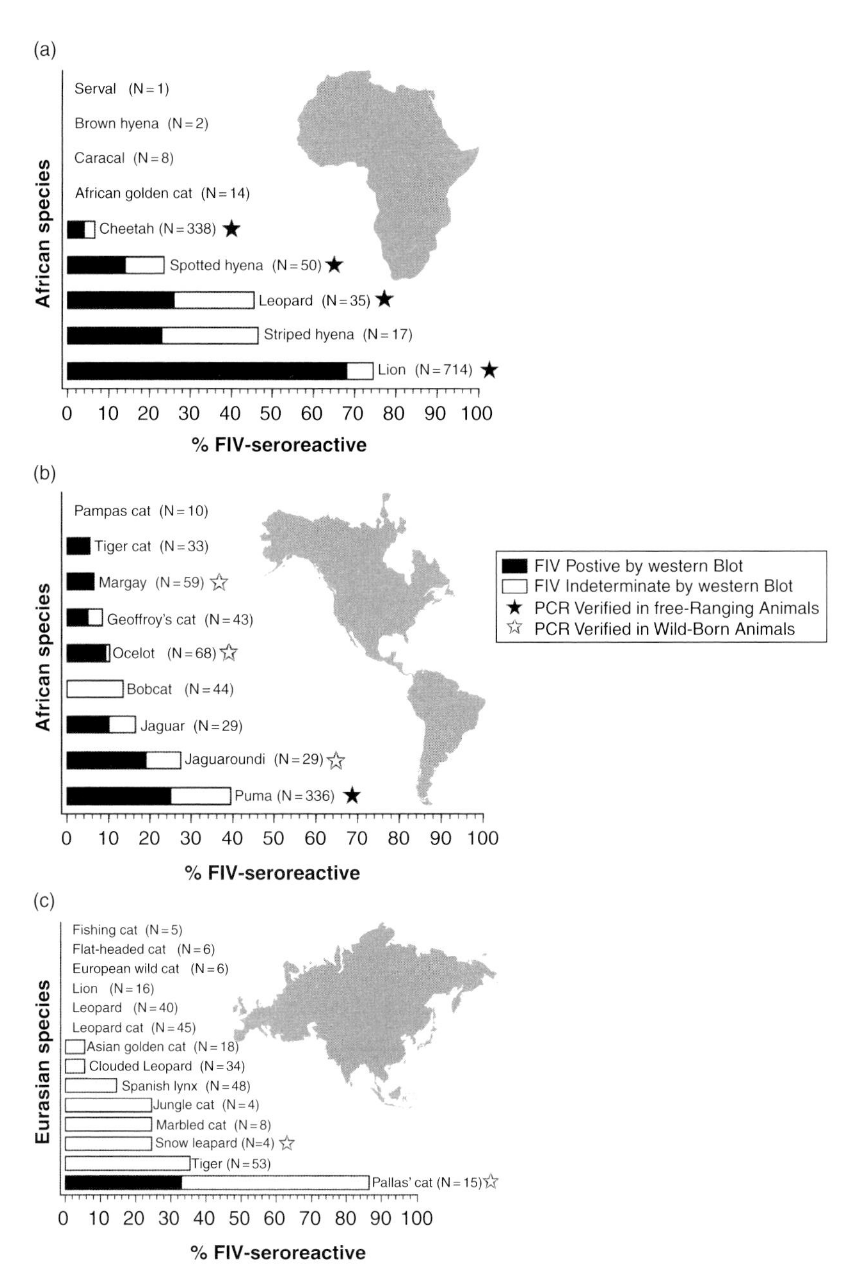

FIGURE 5.3 Prevalence and distribution of FIV (based on antibodies) in free-ranging and captive, but wild-born animals.[24] The dark portion of the bar represents antibody-positives and the white represents indeterminates, giving conservative and liberal estimates of total prevalence of FIV. The number of animals tested for each species is given in parentheses. A black star indicates that PCR verification has been obtained from at least one free-ranging individual for that species. A white star indicates that PCR verification has been obtained from at least one captive, but wild-born individual for that species. a. Percent antibody-reactivity in African carnivores (excluding African golden cat, sand cat, and black footed cat, for which there were no wild-born samples). b. Percent antibody-reactivity in American felids (excluding kodkod and Canadian lynx, for which there were no wild-born samples). c. Percent antibody-reactivity in Eurasian felids (excluding lynx, bay cat, rusty spotted cat and Chinese desert cat, for which there were no wild-born samples).

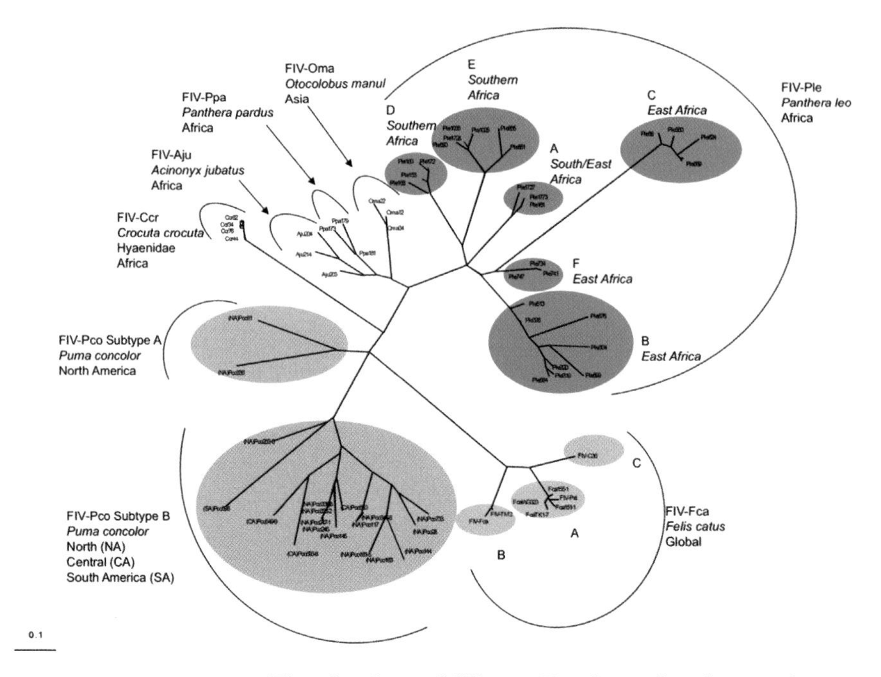

FIGURE 5.4 The relatedness of different FIV viruses from large and small cats species around the world based on maximum likelihood analysis. This is a phylogenetic tree of seventy-two non-identical FIV from seven carnivore species based on a region of pol-RT (420 bp).[31,40] Circles indicate subtypes within FIV_{Ple}, FIV_{Pco} and FIV_{Fca} lineages. [31]

clinical examination among sixty-four free-ranging lions (of which 73 per cent were FIV infected) in East and Southern Africa between 1999 and 2006.[29,30,31] We discovered several examples of AIDS-like pathology including biochemical, clinical and pathogenic manifestations of immune suppression and disease analogous to pathology associated with FIV infection in domestic cats, in AIDS patients and in Simian Immunodeficiency Virus (SIV)-infected macaques. This study[29] provided compelling evidence that FIV contributed to the loss of immune competence in these lions. A contemporary pathogenic study of a wild cohort of SIV-infected chimpanzees also suggested that AIDS-like pathology in that species could occur in the wild African environment where the pathogen burdens and survival stresses were not observed in SIV-infected chimpanzees raised in captivity in Europe and North America.[32,33]

AIDS patients do not actually die of HIV infection alone, rather from their weakened immune system that allows common opportunistic infections, other infections and rare cancers to flourish, whereas in individuals with healthy immune systems these infections are successfully resisted. This raised the question of whether FIV-induced immune deficiency in large cats might contribute to pathology caused by opportunistic infections. An opportunity to inspect this occurred during the mid-1990s in Tanzania when a deadly outbreak of canine distemper virus (CDV; a morbillivirus of dogs) extirpated one-third of the Serengeti lion population over a ten-month interval.[34,35] Because FIV prevalence in East African and Botswana populations approaches 100 per cent in adult lions, perhaps consequent of their highly social behaviour,[26,36,37] the underlying infections of FIV may have had an effect on CDV-induced pathology questioning these great cats.

African Lions harbour six genetically distinct strains, or subtypes, of lion FIV (FIV_{Ple}) resolved by genetic analyses and having distinctive phylogeographic distributions[30,37,38]; three strains were circulating in the Serengeti lions when the outbreak of CDV hit.[35] An epidemiological analysis revealed that infected lions with some genetic types of FIV were twice as likely to survive CDV compared to lions infected with lions who carried alternative strains of FIV.[31,39] The FIV_{Ple} B-associated protective influence was independent of whether individuals were infected with a single strain or with multiple strains. Further, the higher CDV mortality among FIV-carrying individuals actually altered FIV strain incidence, causing a rise in some strains and a decrease in others during the course of the CDV outbreak. This implied that certain FIV strains predisposed carriers of CDV to disease, with similar parallels with FIV strain-specific pathogenicity in domestic cats.[40,41]

As these disease associations were not statistically robust due to the limited number of lions studied (total = 119 lions) our observations should be interpreted cautiously. Nonetheless, the striking influence of FIV on lion immune function clinical disposition, and a potential ancillary role in CDV mortality, affirms that FIV is likely pathogenic in lions. As some strains of FIV_{Ple} are decidedly harmful agents in free-ranging lions, as is FIV_{Fca} in house cats, further scrutiny in the other felidae species afflicted with FIV is required.[29,39] These observations affirm the critical role of certain pathogen/genetic predispositions in outbreaks as well as host genetic influences illustrated in the previous section.

HIV-AIDS: a Modern Human Plague

By far the most revealing data around outbreaks and host pathogen interactions come from human experience. HIV-AIDS represents an important global epidemic, where thirty years of research has revealed much about host control of HIV exposure and infection. Before the first dozen genome-wide association studies were reported to find HIV-infection and AIDS disease susceptibility or resistance genes, single nucleotide polymorphism variants in numerous candidate genes were suggested as being associated with HIV or AIDS in patients.[42,43,44] Over fifty genes have been associated with HIV acquisition or AIDS pathogenesis using large epidemiological cohort studies of people at risk of HIV infection; these are termed AIDS Restriction Genes. The discovery, replication and functional interpretation of AIDS resistance genes has been reviewed in depth and should be consulted to appreciate the details uncovered around each discovery.[42,43,44] One AIDS resistance gene in particular bears brief mention here.

The first and perhaps most provocative ARG involved the description of a gene called *CCR5* in which a 32 base pair deletion frame shift mutation that truncates C-C chemokine receptor 5 (CCR5), one of the key cell surface proteins that HIV uses to enter cells.[45,46,47] People with both alleles being *CCR5-Δ32* have a 100-fold reduction in HIV infection incidence. Thus this homozygous genotype seems to confer near complete protection from HIV infection. In contrast, individuals who have only one copy of this *CCR5-Δ32* allele (heterozygotes) do become infected but, if untreated, have a significant delay in the time it takes to develop full blown AIDS. Since *CCR5-Δ32* genetic resistance was uncovered in 1996, several important translational advances have exploited the discovery of natural genetic resistance to HIV, illustrating the possibilities for important cross-disciplinary research discovery of novel therapeutics (Table 5.3).

First, AIDS therapy, traditionally based on HIV inhibitors, expanded to HIV entry inhibitors that block HIV from entering cells.[48,49,50]

Table 5.3 *Highlights of* CCR5-Δ32 *Discovery and Translation*[44]

CCR5 shown to be HIV entry receptor
Homozygous *CCR5-Δ32* people resist HIV infection
HIV Entry Inhibiters developed and FDA approved for AIDS treatment Fuzeon-T20, Miravirok
CCR5-Δ32 origins dated to historic times and selectively elevated likely by Smallpox, Plague or other scourges
CCR5-Δ32 shown to resist several pathogens
Berlin patient cured of AIDS by *CCR5-Δ32* stem cell transplant
CCR5 human knockouts used for gene therapy for AIDS
CCR5 action implicated in GVHD
Maraviroc CCR5 inhibitor in clinical trials for cancer bone marrow transplants
CCR5-Δ32 carriers shown to be hyper-susceptible to WNV encephalitis.

* Detailed references describing these discoveries are listed in reference 44

These developments were invigorated by the demonstration that *CCR5-Δ32* not only blocked HIV infection but also that the homozygous *CCR5-Δ32* type seemed to have no clinical consequence for uninfected individuals (approximately 2 million *CCR5-Δ32* homozygous carriers exist worldwide). It was clear that this HIV-resistance gene did not arise because of HIV pre-selection, as the resistance genotype was present in the human population long before HIV came along. If that were the case, what event arose to give *CCR5-Δ32* such a relatively high prevalence in European populations? It was a detailed genetic analysis that shed light on the origins of *CCR5-Δ32* during historic time. Findings suggested that this rare mutation was selectively increased in European populations by a devastating epidemic (candidate historical 'Plagues' include the Black-death or Smallpox.[44,51] The next stage in the *CCR5-Δ32* story moved directly to the hospital bedside.

In 2007, Timothy Leigh Brown, dubbed 'The Berlin Patient', was cured of HIV infection when he received a stem cell transplant from a HLA-matched and *CCR5-Δ32/Δ32* homozygous donor. Brown was the first patient in history (40 million patients later and counting) to clear HIV-1 infections and to be pronounced cured of HIV-1 infection.[52,53,54] This amazing medical treatment heralded a series of autologous bone marrow and stem cell transplants to AIDS patients using gene therapy protocols to 'knock out' the patients *CCR5* gene in hopes of replicating the success of the Berlin patient.[55,56] The tentative success of the gene therapy patients plus the Berlin patient fascinated cancer researchers because in both cases there was a marked decrease in 'graft versus host disease (GVHD)', a deadly adverse event that kills 25–30 per cent of heterologous bone marrow transplant patients. Since the AIDS transplant patients seemed free of GVHD, cancer transplant surgeons wondered if *CCR5* might be playing a role in mediating or promoting graft versus host disease. They mounted a cancer-based bone marrow transplant clinical trial (Phase I and II), whereby thirty-eight bone marrow recipients with advanced cancers were treated with a HIV-CCR5 inhibitor to block CCR5 physiological action. Their results showed a dramatic reduction of graft versus host disease without any adverse side effects.[57] If replicated in larger trials, then the *CCR5-Δ32* phenomenon may be a cure for both AIDS and cancer patients in the future.

Since *CCR5-Δ32* was strongly protective against AIDS, the historic search for selective events from infectious agents other than HIV in the distant past began. One approach involved a search for viruses that utilise CCR5 and could be resisted by *CCR5-Δ32* carriers. Notably, CCR5 has been implicated in inflammatory responses to infection by *Mycobacterium tuberculosis* and coxsackievirus, while *CCR5*$^{-/-}$ knockout mice show reduction in survival from *Cryptococcus* and *Listeria* infections. One viral disease showed the opposite effect. A provocative study showed that *CCR5*$^{-/-}$mice were actually hypersensitive to West Nile Virus-WNV pathogenesis and death, prompting a human association study of WNV-infected people.[58,59] West Nile Virus, which was first introduced into New York City in 1999, induces a fatal encephalitis in 20 per cent of human infections. That agent has caused 17,000 infections and 700 encephalitis cases to date, and so far there is no effective vaccine or treatment. A large genetic epidemiological association study of WSV-infected people discovered that *CCR5-Δ32/Δ32* homozygotes were elevated 4–13-fold among the fatal encephalitis cases relative to the survivors.[59] The *CCR5* chemokine gene function apparently exerts a critical defence in the face of WNV. The paradoxical effect of AIDS resistance and WNV hypersensitivity poses a mixed blessing for homozygous carriers of *CCR5-Δ32* and a stark reminder that genetic adaptation can have benefits or even risks when faced with different microbes.

Many other genes have been implicated in HIV-AIDS and in other human disease, but few have progressed so far.[42,44] *CCR5-Δ32* has spawned new therapies, stem cell and gene therapy transplants, cross-over therapies in cancer treatment (Table 5.3) as well as the search for historic disease(s) that could explain its origins. The failure of other AIDS resistance genes to translate can be explained by two aspects. First, even when studies have been reproduced, most gene associations have only modest quantitative effects that diminish enthusiasm for drug design and translation. *CCR5-Δ32*, however, eliminates function effectively; human 'knockout mutations' in most genes are very rare because they eliminate an essential physiological function. Second, even well-accepted associations, such as *HLA*, *KIR*, *APOBEC3g* and *TRIM5å*, do not suggest obvious therapeutic avenues to target. In time, other genes will ultimately develop for other disease applications, but the translation to practical treatments realisation may be slow.

Every debilitating infectious disease or plague we study has heterogeneity in a population of outbred humans and animals. A deep history and undiscovered determinants that regulate who gets infected, who gets sick and who resists the infection are key questions. Now many schools of Public Health worldwide are busy cataloguing heterogeneous individual differences in human epidemics. Most workers now accept that both an understanding of the pathogen and host genetics are critical for full understanding as well as ultimate translation to clinical intervention. The amount of genomic data we are now unravelling from individual humans, vertebrate species and their pathogens is massive, demanding stronger quicker bioinformatic tools to decipher these findings and disease correlations. We are only beginning to uncover the intricate details about deadly diseases which we have been powerless to treat over the millennia. This new era of genomic technology is powerful and informative, but there is a generation ahead of analysis and translation awaiting the curious minds of science. This challenge will be a fascinating and fruitful line of biological and biomedical enquires today and into the future. Let us begin.

Acknowledgements: This research was supported in part by Russian Ministry of Science Mega-grant no.11.G34.31.0068; S.J. O'Brien Principal Investigator.

References

1. Garrett L. (1994) *The Coming Plague*. Penguin: New York.
2. Drazen J.M. (2003) SARS–looking back over the first 100 days. *N Engl J Med* 349:319–20.
3. Drosten C., Preiser W., Gunther S., Schmitz H., Doerr H.W. (2003) Severe acute respiratory syndrome: Identification of the etiological agent. *Trends Mol Med* 9:325–7.
4. Holmes K.V. (2003) SARS-associated coronavirus. *N Engl J Med* 348:1948–51.
5. Enserink M. (2013) SARS: Chronology of the epidemic. *Science* 339:1266–71.
6. Guan Y., Zheng B.J., He Y.Q., Liu X.L., Zhuang Z.X., Cheung C.L., Luo S.W., Li P.H., Zhang L.J., Guan Y.J., Butt K.M., Wong K.L., Chan K.W., Lim W., Shortridge K.F., Yuen K.Y., Peiris J.S., Poon L.L. (2003) Isolation and characterization of viruses related to the SARS coronavirus from animals in southern China. *Science* 302:276–8.
7. Lau S.K., Woo P.C., Li K.S., Huang Y., Tsoi H.W., Wong B.H., Wong S.S., Leung S.Y., Chan K.H., Yuen K.Y. (2005) Severe acute respiratory syndrome coronavirus-like virus in Chinese horseshoe bats. *Proc Natl Acad Sci USA* 102:14040–5.
8. Li W., Shi Z., Yu M., Ren W., Smith C., Epstein J.H., Wang H., Crameri G., Hu Z., Zhang H., Zhang J., McEachern J., Field H., Daszak P., Eaton B.T., Zhang S., Wang L.F. (2005) Bats are natural reservoirs of SARS-like coronaviruses. *Science* 310:676–9.
9. Poon L.L., Chu D.K., Chan K.H., Wong O.K., Ellis T.M., Leung Y.H., Lau S.K., Woo P.C., Suen K.Y., Yuen K.Y., Guan Y., Peiris J.S. (2005) Identification of a novel coronavirus in bats. *J Virol* 79:2001–9.
10. Savarino A. (2005) Expanding the frontiers of existing antiviral drugs: possible effects of HIV-1 protease inhibitors against SARS and avian influenza. *J Clin Virol* 34:170–8.
11. Wei P., Fan K., Chen H., Ma L., Huang C., Tan L., Xi D., Li C., Liu Y., Cao A., Lai L. (2006) The N-terminal octapeptide acts as a dimerization inhibitor of SARS coronavirus 3C-like proteinase. *Biochem Biophys Res Commun* 339:865–72.
12. Foley J.E., Poland A., Carlson J., Pedersen N.C. (1997) Risk factors for feline infectious peritonitis among cats in multiple-cat environments with endemic feline enteric coronavirus. *J Am Vet Med Assoc* 210:1313–18.
13. Ballesteros M.L., Sanchez C.M., Enjuanes L. (1997) Two amino acid changes at the N-terminus of transmissible gastroenteritis

coronavirus spike protein result in the loss of enteric tropism. *Virology* 227:378–88.
14. Sanchez C.M., Izeta A., Sanchez-Morgado J.M., Alonso S., Sola I., Balasch M., Plana-Duran J., Enjuanes L. (1999) Targeted recombination demonstrates that the spike gene of transmissible gastroenteritis coronavirus is a determinant of its enteric tropism and virulence. *J Virol* 73:7607–18.
15. Heeney J.L., Evermann J.F., McKeirnan A.J., Marker-Kraus L., Roelke M.E., Bush M., Wildt D.E., Meltzer D.G., Colly L., Lukas J., et al. (1990) Prevalence and implications of feline coronavirus infections of captive and free-ranging cheetahs (*Acinonyx jubatus*). *J Virol* 64:1964–72.
16. O'Brien S.J., Roelke M.E., Marker L., Newman A., Winkler C.A., Meltzer D., Colly L., Evermann J.F., Bush M., Wildt D.E. (1985) Genetic basis for species vulnerability in the cheetah. *Science* 227:1428–34.
17. Evermann J.F., Heeney J.L., Roelke M.E., McKeirnan A.J., O'Brien S.J. (1988) Biological and pathological consequences of feline infectious peritonitis virus infection in the cheetah. *Arch Virol* 102:155–71.
18. Pearks Wilkerson A.J., Teeling E.C., Troyer J.L., Bar-Gal G.K., Roelke M., Marker L., Pecon-Slattery J., O'Brien S.J. (2004) Coronavirus outbreak in cheetahs: Lessons for SARS. *Curr Biol* 14: R227–28.
19. Menotti-Raymond M., O'Brien S.J. (1993) Dating the genetic bottleneck of the African cheetah. *Proc Natl Acad Sci USA* 90:3172–6.
20. Yuhki N., Beck T., Stephens R.M., Nishigaki Y., Newmann K., O'Brien S.J. (2003) Comparative genome organization of human, murine, and feline MHC class II region. *Genome Res* 13:1169–79.
21. Yuhki N., O'Brien S.J. (1990) DNA variation of the mammalian major histocompatibility complex reflects genomic diversity and population history. *Proc Natl Acad Sci USA* 87:836–40.
22. CDC (1981). MMWR Kaposi's sarcoma and pneumocystis pneumonia among homosexual men – New York City and California – Los Angeles. *MMWR Morb Mortal Wkly Rep* 1981, 30:250–2; ibid. 305–8.
23. Pedersen N.C., Ho E.W., Brown M.L., Yamamoto J.K. (1987) Isolation of a T-lymphotropic virus from domestic cats with an immunodeficiency-like syndrome. *Science* 235:790–3.
24. Troyer J.L., Pecon-Slattery J., Roelke M.E., Johnson W., VandeWoude S., Vazquez-Salat N., Brown M., Frank L.,

Woodroffe R., Winterbach C., Winterbach H., Hemson G., Bush M., Alexander K.A., Revilla E., O'Brien S.J. (2005) Seroprevalence and genomic divergence of circulating strains of feline immunodeficiency virus among Felidae and Hyaenidae species. *J Virol* 79:8282–94.
25. Olmsted R.A., Langley R., Roelke M.E., Goeken R.M., Adger-Johnson D., Goff J.P., Albert J.P., Packer C., Laurenson M.K., Caro T.M., et al. (1992) Worldwide prevalence of lentivirus infection in wild feline species: Epidemiologic and phylogenetic aspects. *J Virol* 66:6008–18.
26. Carpenter M.A., O'Brien S.J. (1995) Coadaptation and immunodeficiency virus: Lessons from the Felidae. *Curr Opin Genet Dev* 5:739–45.
27. Antunes A., Troyer J.L., Roelke M.E., Pecon-Slattery J., Packer C., Winterbach C., Winterbach H., Hemson G., Frank L., Stander P., et al. (2008) The evolutionary dynamics of the lion Panthera leo revealed by host and viral population genomics. *PLoS Genet* 4:e1000251.
28. Packer C., Altizer S., Appel M., Brown E., Martenson J., O'Brien S.J., Roelke-Parker M., Hofmann-Lehmann R., Lutz H. (1999) Viruses of the Serengeti: Patterns of infection and mortality in African lions. *J Anim Ecol* 68:1161–78.
29. Roelke M.E., Brown M.A., Troyer J.L., Winterbach H., Winterbach C., Hemson G., Smith D., Johnson R.C., Pecon-Slattery J., Roca A.L., et al. (2009) Pathological manifestations of feline immunodeficiency virus (FIV) infection in wild African lions. *Virology* 390:1–12.
30. Pecon-Slattery J., Troyer, J.L., Johnson, W.E. O'Brien, S.J.: Evolution of feline immunodeficiency virus in Felidae: implications for human health and wildlife ecology. *Vet Immunol Immunopathol* 123:32–44, 2008.
31. O'Brien Stephen J., Troyer Jennifer L., Brown Meredith A., Johnson Warren E., Antunes Agostinho, Roelke Melody E., Pecon-Slattery Jill (2012) Emerging Viruses in the Felidae: Shifting Paradigms *Viruses* 4:236–57.
32. Keele B.F., Jones J.H., Terio K.A., Estes J.D., Rudicell R.S., Wilson M.L., Li Y., Learn G.H., Beasley T.M., Schumacher-Stankey J., et al. (2009) Increased mortality and AIDS-like immunopathology in wild chimpanzees infected with SIVcpz. *Nature* 460:515–19.
33. Weiss R.A., Heeney J.L. (2009) Infectious Diseases,; An ill wind for chimps. *Nature* 460:470–1.
34. Carpenter M.A., Appel M.J., Roelke-Parker M.E., Munson L., Hofer H., East M., O'Brien S.J. (1998) Genetic characterization of

canine distemper virus in Serengeti carnivores. *Vet Immunol Immunopathol* 65:259–66.
35. Roelke-Parker M.E., Munson L., Packer C., Kock R., Cleaveland S., Carpenter M., O'Brien S.J., Pospischil A., Hofmann-Lehmann R., Lutz H., et al. (1996) A canine distemper virus epidemic in Serengeti lions (Panthera leo). *Nature* 379:441–5.
36. Brown E.W., Yuhki N., Packer C., O'Brien S.J. (1994) A lion lentivirus related to feline immunodeficiency virus: epidemiologic and phylogenetic aspects. *J Virol* 68:5953–68.
37. Troyer J.L., Pecon-Slattery J., Roelke M.E., Black L., Packer C., O'Brien S.J. (2004) Patterns of feline immunodeficiency virus multiple infection and genome divergence in a free-ranging population of African lions. *J Virol* 78:3777–91.
38. Brown E.W., Miththapala S., O'Brien S.J. (1993) Prevalence of exposure to feline immunodeficiency virus in exotic felid species. *J Zoo Wildl Med* 24:357–64.
39. Troyer J.L., Roelke M.E., Jespersen J.M., Baggett N., Buckley-Beason V., Macnulty D., Craft M., Packer C., Pecon-Slattery J., O'Brien S.J. (2011) FIV diversity: FIV(Ple) subtype composition may influence disease outcome in African lions. *Vet Immunol Immunopathol* 143:338–46.
40. Pecon-Slattery J., Troyer J.L., Johnson W.E., O'Brien S.J. (2008) Evolution of feline immunodeficiency virus in Felidae: Implications for human health and wildlife ecology. *Vet Immunol Immunopathol* 123:32–44.
41. VandeWoude S., Hageman C.L., Hoover, E.A. (2002) Domestic cats infected with lion or puma lentivirus develop anti-feline immunodeficiency virus immune responses. *J Acquir Immune Defic Syndr* 2003(34):20–31.
42. An P., Winkler C.A. (2010) Host genes associated with HIV/AIDS: advances in gene discovery. *Trends Genet* 26:119–31.
43. O'Brien S.J., Nelson G.W. (2004) Human genes that limit AIDS. *Nat Genet* 36:565–74.
44. O'Brien S.J., Hendrickson S. (2013) Host genomic influences on HIV/ AIDS. *Genome Biology* 14:201.
45. Dean M., Carrington M., Winkler C., Huttley G.A., Smith M.W., Allikmets R., Goedert J.J., Buchbinder S.P., Vittinghoff E., Gomperts E., Donfield S., Vlahov D., Kaslow R., Saah A., Rinaldo C., Detels R., O'Brien S.J. (1996) Genetic restriction of HIV-1 infection and progression to AIDS by a deletion allele of the CKR5 structural gene. *Science* 273:1856–62.

46. Liu R., Paxton W.A., Choe S., Ceradini D., Martin S.R., Horuk R., MacDonald M.E., Stuhlmann H., Koup R.A., Landau N.R. (1996) Homozygous defect in HIV-1 coreceptor accounts for resistance of some multiply-exposed individuals to HIV-1 infection. *Cell* 86:367–77.
47. Samson M., Libert F., Doranz B.J., Rucker J., Liesnard C., Farber C.M., Saragosti S., Lapoumeroulie C., Cognaux J., Forceille C., Muyldermans G., Verhofstede C., Burtonboy G., Georges M., Imai T., Rana S., Yi Y., Smyth R.J., Collman R.G., Doms R.W., Vassart G., Parmentier M. (1996) Resistance to HIV-1 infection in Caucasian individuals bearing mutant alleles of the CCR-5 chemokine receptor gene. *Nature* 382:722–5.
48. Aquaro S., D'Arrigo R., Svicher V., Perri G.D., Caputo S.L., Visco-Comandini U., Santoro M., Bertoli A., Mazzotta F., Bonora S., Tozzi V., Bellagamba R., Zaccarelli M., Narciso P., Antinori A., Perno C.F. (2006) Specific mutations in HIV-1 gp41 are associated with immunological success in HIV-1-infected patients receiving enfuvirtide treatment. *J Antimicrob Chemother* 58:714–22.
49. Dorr P., Westby M., Dobbs S., Griffin P., Irvine B., Macartney M., Mori J., Rickett G., Smith-Burchnell C., Napier C., Webster R., Armour D., Price D., Stammen B., Wood A., Perros M. (2005) Maraviroc (UK-427,857), a potent, orally bioavailable, and selective small-molecule inhibitor of chemokine receptor CCR5 with broad-spectrum anti-human immunodeficiency virus type 1 activity. *Antimicrob Agents Chemother* 49:4721–32.
50. Manfredi R., Sabbatani S. (2006) A novel antiretroviral class (fusion inhibitors) in the management of HIV infection: Present features and future perspectives of enfuvirtide (T-20). *Curr Med Chem J* 13:2369–84.
51. Hedrick Philip W., Verrelli Brian C. (2006) 'Ground Truth' for selection on CCR5-Δ32. *Trends Genet* 22:293–6.
52. Allers K., Hütter G., Hofmann J., Loddenkemper C., Rieger K., Thiel E., Schneider T. (2011) Evidence for the cure of HIV infection by CCR5Δ32/Δ32 stem cell transplantation. *Blood* 117:2791–9.
53. Cohen J. (2011) The emerging race to cure HIV infections. *Science* 332:784–9.
54. Hutter G., Nowak D., Mossner M., Ganepola S., Mussig A., Allers A., Schneider T., Hofmann J., Kucherer C, Blau O., Blau I., Hofmann W.K., Thiel E. (2009) Long-term control of HIV by CCR5 delta32/delta32 stem-cell transplantation. *New Engl J Med* 360:692–8.
55. June C., Levine B. (2012) Blocking HIV's attack. *Sci Am* 3:54–9.

56. Tebas P., Stein D., Tang W.W., Frank I., Wang S.Q., Lee G., Spratt S.K., Surosky R.T., Giedlin M.A., Nichol G., Holmes M.C., Gregory P.D., Ando D.G., Kalos M., Collman R.G., Binder-Scholl G., Plesa G., Hwang W.T., Levine B.L., June C.H. (2014) Gene editing of CCR5 in autologous CD4 T cells of persons infected with HIV. *N Engl J Med.* Mar 6;370(10):901–10. doi: 10.1056/NEJMoa1300662.
57. Reshef R, Luger SM, Hexner EO, Loren AW, Frey NV, Nasta SD, Goldstein SC,Stadtmauer EA, Smith J, Bailey S, Mick R, Heitjan DF, Emerson SG, Hoxie JA, Vonderheide RH, Porter DL: Blockade of lymphocyte chemotaxis in visceral graft-versus-host disease. N Engl J Med 2012, 367:135–145.
58. Glass W.G., Lim J.K., Cholera R., Pletnev A.G., Gao J.L., Murphy P.M. (2005) Chemokine receptor CCR5 promotes leukocyte trafficking to the brain and survival in West Nile virus infection. *J Exp Med* 202:1087–98.
59. Glass W.G., McDermott D.H., Lim J.K., Lekhong S., Yu S.F., Frank W.A., Pape J., Cheshier R.C., Murphy P.M. (2006) CCR5 deficiency increases risk of symptomatic West Nile virus infection. *J Exp Med.* 2006 Jan 23;203(1):35–40.
60. Willett, B.J., Flynn, J.N., and Hosie, M.J. 1997. FIV infection of the domestic cat: an animal model for AIDS. Immunol. Today 18:182–189.

6 Plagues and Socioeconomic Collapse

IAN MORRIS

'Civilization both in East and West was visited by a destructive plague that devastated nations and caused populations to vanish', the Arab traveller and philosopher Ibn Khaldûn wrote in 1377. 'It swallowed up many of the good things of civilization and wiped them out.'[1]

Ibn Khaldûn lived through the Black Death that had begun ravaging the Middle East in 1346, and clearly knew what he was talking about. By 1377, the plague had killed between one-third and one-half of the people in China, the Middle East and Europe (although it seems scarcely to have touched Japan, Southeast Asia, India, sub-Saharan Africa or the Americas).[2] 'As far as can be found in written records', the Florentine chronicler Matteo Villani concluded, 'there has been no more widespread judgment by mortal illness from the universal Deluge to the present, nor one that embraced more of the universe, than the one that has occurred in our own day'.[3]

And yet, nearly seven centuries later, humanity has not just survived the Black Death; it has positively flourished in its wake. There are now roughly seventeen times as many of us on the planet as there were in 1346. Each of us, on average, lives roughly twice as long as our fourteenth-century ancestors and produces, on average, fifteen times as much wealth per year.[4] We live in an age of abundance that would have seemed like a magical kingdom to the people who endured the Black Death.

Any lecture series on plagues would need to ask how and why this happened, and when the Master and Fellows of Darwin College invited me to come back to Cambridge and suggest some answers, I jumped at the chance.[5] Ernest Gellner's Darwin Lecture on the origins of society, delivered in the very first series (back in 1986),[6] permanently changed the way I think about the past, and I could not possibly turn down this extraordinary honour. That said, the invitation was also intimidating, not least because the way I like to look at such big questions – through the lens of long-term history – has hardly been very prominent in previous

Darwin Lectures. On reflection, however, that seemed to me all the more reason to accept. Thinking about long-term history has turned me into much more of a Darwinist than I was when I listened to Professor Gellner's lecture almost thirty years ago,[7] and perhaps my own paper can nudge other historians towards the same Darwinian light.

In this chapter, I take an approach that is at once broad and narrow. It is narrow because rather than trying to explain everything about the historical consequences of plagues, and probably ending up explaining nothing, I zoom in on a single dimension; but it is broad because the only way to do that is by comparing multiple plagues, which means working on the biggest possible scales, both chronologically and geographically. When we do this, I suggest, we get some added bonuses. Once we have identified the broad patterns of the past, we can also search for both the forces that have driven them and the forces that derail them; and if we can identify these, there is a chance that we can also project the long-term trends forward to get some idea of what the twenty-first century might have in store.

The question I ask is whether plagues cause socioeconomic collapse, and the answer I offer is a resounding yes and no. As so often, the answer depends largely on what we take words to mean. History seems to show that some kinds of plagues are a sufficient and perhaps even necessary cause for some kinds of socioeconomic collapse, but not for other kinds. We might say that plagues function as shocks – sometimes massive ones – to socioeconomic systems. They force people to respond. Sometimes the response has disastrous consequences; sometimes, it ends up leaving the society and economy stronger than it had been before the shock; and sometimes the shock is so severe that it simply overwhelms people's capacity to respond at all. Depending on how the details turn out, the coming century might, or might not, see a major new plague; and that might, or might not, bring one or more kinds of socioeconomic collapse in its wake. All things are possible – although, for reasons I will return to at the end of the chapter, I think that while the first of these catastrophes (plague) is quite likely in the coming century, only certain kinds of the second (collapse) are very probable.

My plan for the rest of the chapter is straightforward. First, I will look at the socioeconomic consequences of the fourteenth-century Black Death, which has been intensively studied. In the second section, I will look at several other less well-known plagues; in the third, I will put the results together to ask what we can say about the relationship between plague and socioeconomic collapse; and in the final section, I will ask what history can tell us about plague and collapse in the century to come.

The Black Death and Socioeconomic Collapse

Before we can decide whether plagues cause socioeconomic collapse, we of course need to decide what we mean by these words. 'Plague' I will postpone to the next section, because the Black Death is for most historians not just a plague but *the* plague, the archetype to which we must compare all contenders for the label 'plague'.[8] 'Socioeconomic collapse', however, demands immediate attention.

Unfortunately, the scholarly literature is determinedly vague. According to the archaeologist Joseph Tainter, whose 1988 book *The Collapse of Complex Societies* is often (rightly) seen as the founding document of modern collapse studies, '*A society has collapsed when it displays a rapid, significant loss of an established level of sociopolitical complexity*', while the biologist and geographer Jared Diamond, whose book *Collapse* must be the most widely read work on the subject, says, 'By collapse, I mean a drastic decrease in human population size and/or political/economic/social complexity, over a considerable area, for a considerable time.'[9] The approach I take here, however, calls for a little more precision, and so I want to make two kinds of distinction.

The first concerns the scale of collapse, and to spell out what I mean, I will fall back on a way of thinking that I pursued in earlier work in a pair of books called *Why the West Rules – For Now* and *The Measure of Civilisation.*[10] In these I tried to quantify what I called 'social development', by which I basically meant societies' abilities to get things done in the world – or, to put it more formally,

> the bundle of technological, subsistence, organizational, and cultural accomplishments through which people feed, clothe, house, and reproduce

> themselves, explain the world around them, resolve disputes within their communities, extend their power at expense of other communities, and defend themselves against others' attempts to extend power.[11]

There are many ways we might construct such an index, but I took my lead from the United Nations Development Programme, which, since 1990, has published an annual human development index.[12] To calculate scores, the UN economists break down human development into discrete quantifiable traits, and then reduce these to the smallest possible set (life expectancy at birth, lifetime real wages and education) that more or less covers the larger concept of 'human development' (basically, how well a government is doing at creating conditions that allow its citizens to realise their innate human potential).[13] Since 'social development' is a different concept from 'human development', I needed to use different traits (energy capture per person, organisational capacity [measured through the proxy of settlement size], information technology and war-making capacity), but I convinced myself, at least, that these gave me a tool not only for measuring social development over the long run (going back, in fact, to the final stages of the last ice age, around 14,000 BC) but also comparing development in Eastern and Western Eurasia.[14]

This social development index allows us to plug some numbers into Tainter's and Diamond's looser definitions of socioeconomic collapse. We can then choose whether to set an arbitrary definition of a collapse (say, a 10 per cent fall in social development scores) or just examine raw scores, but either way, the index suggests that the Black Death correlated with a serious socioeconomic collapse. In East and West alike, development scores fell by 11 per cent between 1300 and 1400 (Figure 6.1), and the historical details suggest that the correlation was causal.[15]

Immediately, though, problems begin to arise. Figure 6.1, looking at the years between 800 and 1400, shows that the Black Death correlated with socioeconomic collapse; but Figure 6.2, extending the story forward to 1800, can be interpreted as showing just the opposite. Between 1400 and 1800, development scores soared faster than in any earlier four-century period, by 39 per cent in the East and 71 per cent in the West. This suggests that as well as defining 'collapse' along the *y* axis of

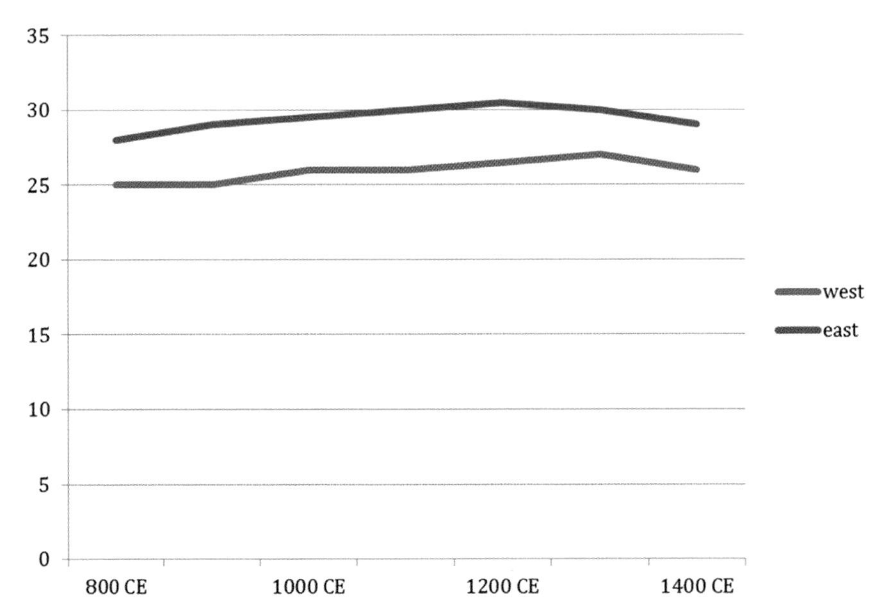

FIGURE 6.1 Eastern and Western social development scores, 800–1400 (data from Morris 2013).

these graphs, we also need to define it along the *x* axis, asking how long an episode of decline and fall must last to count as a proper collapse. In what follows, I will divide collapses into three classes – what I will call the single-century collapse (i.e., social development scores are at least 10 per cent lower a century after the plague than they were before it), the multi-century collapse (i.e., scores remain lower for hundreds of years), and the permanent collapse (i.e., development scores never regain the level they were at before the plague).

The Black Death, obviously, was a single-century collapse. We might even see Figure 6.2 as meaning that despite the horrors it brought to the people who suffered from it, the plague was a good thing in the long run, because the post-plague regeneration was so much bigger than the initial collapse. This is in fact much the line that mainstream British, French and American historians have taken since at least the 1950s about Europe's experience. The Black Death, this theory runs, destabilised medieval European society by undermining institutions, weakening

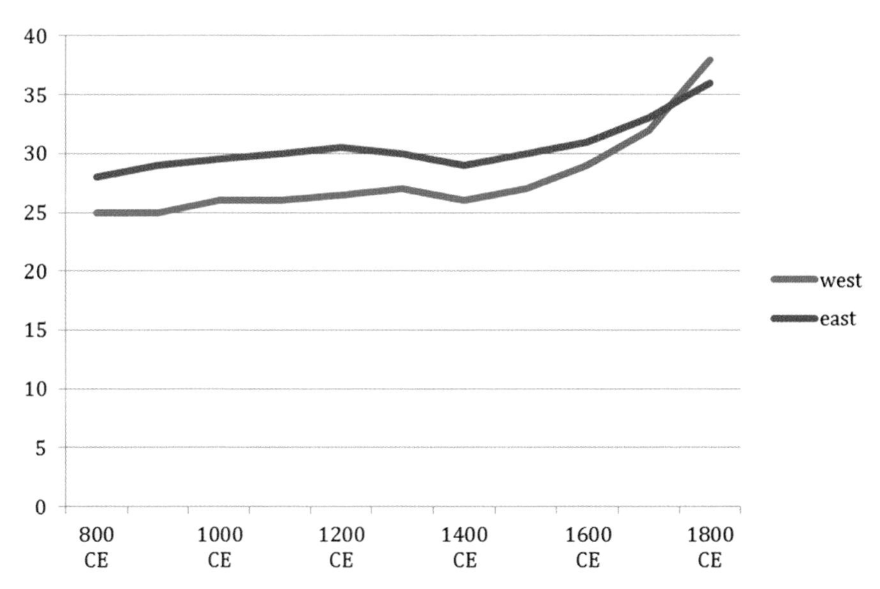

FIGURE 6.2 Eastern and Western social development scores, 800–1800 (data from Morris 2013).

bonds of personal dependence and unleashing violent struggles. Feudalism broke down, and capitalism filled the void. Population was growing again by 1500, but by that point the new social relationships were taking hold, producing socioeconomic take-off and ultimately the industrial revolution.[16]

Marxist historians, however, have challenged this theory's demographic determinism.[17] In the most influential account, Robert Brenner argued that if we look at Europe as a whole, rather than separate kingdoms, we see that the real story was much more complicated. 'Different outcomes', he concluded, 'proceeded from similar demographic trends at different times and in different areas of Europe ... [T]he same ... pressure of population could, and did, lead to changes in the distribution of income favorable to the lords *or* to the peasants – opposite outcomes – depending on the social-property relationships and balances of class forces.'[18] East of the river Elbe, Brenner suggested, the balance of class forces favoured the landlords, who imposed a 'second feudalism' on the peasantry after 1500;[19] but West of the Elbe, it favoured the poor, who escaped from serfdom. As Brenner saw it, the plague and its demographic consequences were important, and did drive collapse followed by regeneration; but what really drove collapse and regeneration was the class struggle.

Heated debates ensued in the 1970s–80s,[20] but since the late 1990s economic historians have added a new wrinkle to argument. Years of painstaking work in the archives have yielded serial data on real wages, stretching back in some parts of Europe to the thirteenth century.[21] This evidence reveals just what Malthus and Ricardo would have expected: on both sides of the Elbe, real wages roughly doubled between 1350 and 1450 because the plague shifted land:labour ratios in the workers' favour (Figure 6.3), only for the process to go into reverse after 1500 as rising population shifted the land:labour ratio back again. By 1600, real wages in southern and eastern Europe were generally back where they had been around 1300. Only in northwest Europe did wages remain high, for important reasons that I will return to below. Sixteenth- and seventeenth-century European peasants rightly looked back on the fifteenth century as a golden age of cakes and ale. 'In the past', a German traveller noted in 1560, 'they ate differently at the peasant's house. Then, there was meat and food in profusion'. Today, though, 'everything is truly changed . . . the food of the most comfortably off peasants is almost worse than that of day labourers and valets in the old days'.[22]

The wage data are less good outside Europe,[23] but the economic historians Sevket Pamuk and Maya Shatzmiller have suggested that incomes rose by 50 per cent in fourteenth-century Cairo. In the late fifteenth century, workers in Istanbul were paid even better, although wages then fell sharply in the sixteenth century (Figure 6.4).[24] The Middle Eastern pattern is messier than the European (wages were already rising in Cairo in the thirteenth century, but then fell in the late fourteenth before reviving in the fifteenth), but seems consistent with it. Still further east, in China, we currently have no numbers at all,[25] although the qualitative evidence certainly fits the European fifteenth-century boom/sixteenth-century bust pattern. Looking back in 1609, a county prefect near Nanjing lamented that in the olden days, 'Every family was self-sufficient with a house to live in, land to cultivate, hills

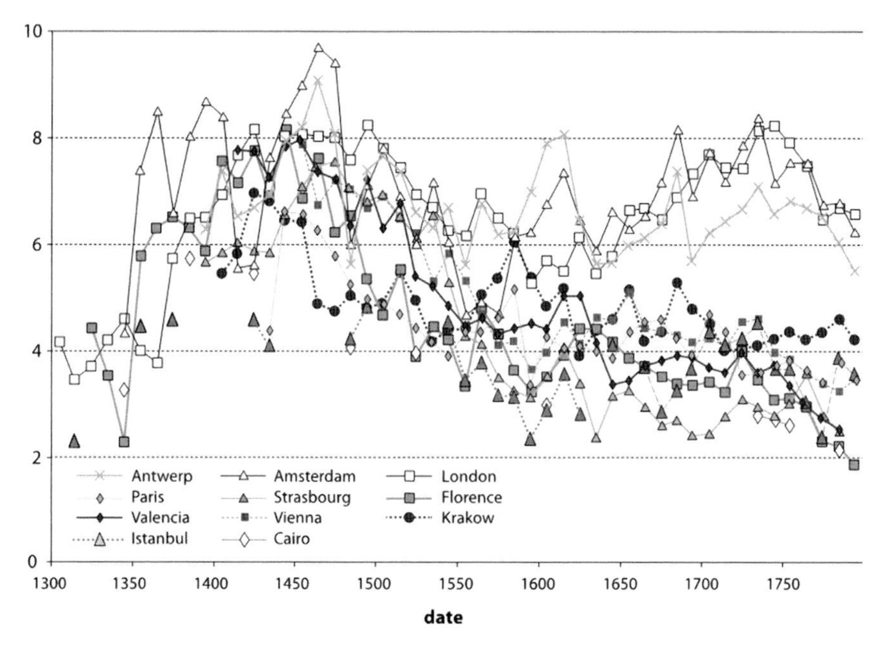

FIGURE 6.3 Real wages in selected European and Mediterranean cities, 1300–1800 (after Pamuk 2007).

from which to cut firewood, gardens in which to grow vegetables.' Now, though, 'nine out of ten are impoverished . . . Avarice is without limit, flesh injures bone . . . Alas!'[26]

So did the Black Death cause socioeconomic collapse? The social development, Marxist and real-wage approaches give different answers. Yes, say the social development scores and mainstream historiography on Europe, with the proviso that the collapse was of the single-century kind, followed by a strong regeneration. From this perspective, the plague was bad in the short run but good in the long run. No, says Brenner and many other Marxists; the Black Death was important, but what determined the depth, length and consequences of collapse and regeneration was class structures. Yes, say the trends in real wages, but again with a proviso: for the poor, the plague was good in the short run, but across most of Europe made little difference in the long run.

We seem no nearer to answering the question.

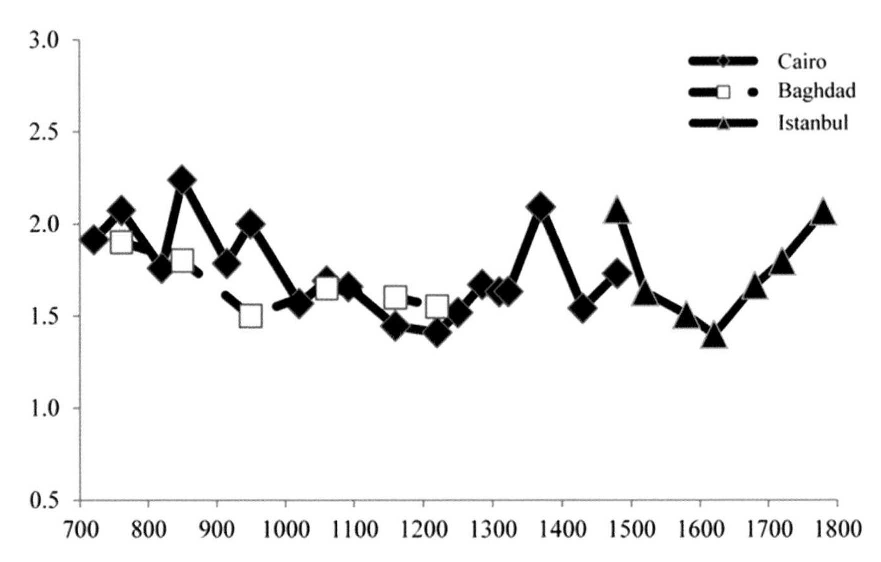

FIGURE 6.4 Real wages in Cairo, Baghdad and Istanbul, 700–1800 (after Pamuk and Shatzmiller 2014).

Other Plagues and Socioeconomic Collapse

One possible way forward is by out-Brennering Brenner, comparing the consequences of plagues not just across the whole of Europe since the fourteenth century but across the whole world in all ages. The inspiration for anyone who tries this is of course William McNeill's *Plagues and Peoples*,[27] the extraordinary book that founded the entire field of comparative studies of the history of disease.

McNeill provided a survey of what could be said about this subject in the mid-1970s, leading to the insight that humanity has always been 'caught in a precarious equilibrium between the microparasitism of disease organisms and the macroparasitism of large-bodied predators, chief among which have been other human beings'.[28] This, however, brings us up against a new challenge, of how to define 'plague' itself. My sense is that epidemiologists normally do so narrowly, applying the word solely to bacterial infections caused by *Yersinia pestis*, while historians prefer broad definitions like McNeill's. As always, there is a trade-off. Loose definitions let us generalise broadly, but run the risk of descending into vagueness; tight definitions provide precision, but risk producing categories with only one case in them. When preparing my Darwin lecture, I assumed (although I did not bother to ask) that the organisers wanted to know whether pandemic infectious diseases in general (rather than *Y. pestis* in particular) cause collapse, and so I will define 'plague' as any contagious pandemic that causes high mortality. How many deaths count as 'high mortality' is as much a judgement call as how much socioeconomic decline counts as 'collapse', and once again analysts are free either to set a threshold (say, excess mortality of 10 per cent within a particular continent, or 5 per cent globally) or to deal in raw numbers, measuring the 'plagueness' of specific pandemics on a sliding scale.

A 10 per cent (continental)/5 per cent (global) cut-off would set the bar high. The Black Death easily clears it, killing somewhere between a third and a half of the people in Europe, the Middle East and China between 1350 and 1400, and probably cutting world population by more than 10 per cent. No modern epidemic, however, seems to have reached this level. The bubonic plague that ravaged India after 1896 killed perhaps 5 per cent of the population of 250 million, and even the greatest killer in history – the 1918–19 H1N1 influenza epidemic – falls short of the cut-off. It ended roughly 50 million lives; but there were so many people in the world a century ago that the flu's victims add up to just 3 per cent of the total. Even in Europe, where 30 million may have died, the death rate was well under 10 per cent.[29] The AIDS epidemic, which has inflicted such suffering (especially in sub-Saharan Africa) since the 1980s, had even less demographic impact: its death toll – around 36 million by the end of 2013 – represents just 0.5 per cent of the world's population.[30]

The most severe plague before the Black Death was probably an outbreak that Roman historians call the Antonine Plague, which seems to have erupted simultaneously in Syria and on China's northwest frontier in the year 161. Scientists have not yet pinpointed the pathogens responsible, but eyewitness accounts make the disease sound a lot like smallpox. According to He Gong, a Chinese doctor of the early fourth century,

> epidemic sores ... attack the head, face and trunk. In a short time, these sores spread all over the body. They have the appearance of hot boils

> containing some white matter. While some of these pustules are drying up a fresh crop appears. If not treated early the patients usually die. Those who recover are disfigured by purplish scars.

In the late second century, the disease killed perhaps a quarter of the people in Roman Egypt and unknown, but certainly huge, numbers in the rest of the empire and in China.[32] Nor was that the end of the suffering: the disease kept coming back, roughly once every generation, for the next century and a half. For a while around 250, we are told, five thousand people were dying every day in the city of Rome. Italian archaeologists have recently excavated a massive plague cemetery from these years in Egypt, although they did not succeed in extracting DNA.[33] In China, the worst years came between 310 and 322.[34] Globally, the death toll might well have surpassed 5 per cent, although at this point we can only speculate. My own speculation is that the plague did clear the 10 per cent/5 per cent cut-off, but by a much smaller margin than the Black Death.

Much remains obscure about the Antonine Plague, but McNeill was probably right to suggest that it evolved, like the Black Death, when the previously distinct disease pools of western and eastern Eurasia merged.[35] The Black Death came just a few generations after the Mongols had established loose rule over the steppes that link the two ends of Eurasia, greatly facilitating East–West movement, and the Antonine Plague broke out right around the time of the first convincingly documented (by mitochondrial DNA[36] as well as texts) direct contacts between Rome and China, in an era of sharply accelerating movements across the steppes.[37] If this theory is correct, the common view of the Antonine Plague and Black Death as exogenous shocks to the agrarian empires is wrong; rather than bursting in from the outside, the plagues were created by the expansion of the system as a whole, which produced greater human mobility and a new disease environment.[38]

The evidence is patchy, but in his most recent analysis of the question, the economic historian and demographer Walter Scheidel concludes that in Egypt (thanks to the survival of papyri, the best documented region) 'the Antonine Plague had a non-trivial impact on the real income of rural wage laborers and farm tenants'. He chooses these words carefully, because 'comparison with the sheer scale of post-plague changes in real wages in later periods of Egyptian history as well as in late medieval England and other parts of Europe speaks against the notion that the Antonine Plague was a truly devastating crisis'.[39]

The long-term socioeconomic impact of the Antonine Plague was also very different from that of the Black Death.[40] The second and fourteenth centuries both saw sharp downturns in social development, but while scores recovered rapidly after 1400, after 200 they had continued dropping. By 400, Chinese development had fallen 10 per cent from its first-century level, and by 500 Roman development had tumbled a terrifying 25 per cent (Figure 6.5). This was a multi-century socioeconomic collapse.

Social development scores moved in parallel at the two ends of Eurasia for several hundred years, but in the fifth century the trajectories diverged. In the East, decline and fall bottomed out before 500, and before 600 development had regained its first-century level. In the West, however, the fall in development only accelerated. Between 500 and 800, Western scores fell by a further 18 per cent, while Eastern scores rose by 10 per cent. Overall, Western development scores fell by 34 per cent between 100 and 700, the biggest decline in any period of this length since 14,000 BC.

Plague clearly had a lot to do with this great East–West divergence. In 541, as the Byzantine emperor Justinian's attempts to reunite the Roman Empire reached their height, a nasty new illness was reported in Egypt. People broke out in fevers and their groins and armpits swelled. After a day or two, the swellings would blacken. Patients fell into comas and delirium, and within another day or two most died in agony. The sickness – recently confirmed as bubonic plague by DNA analyses of burials in Germany[41] – reached Constantinople in 542, reportedly killing 100,000 people.[42]

The Justinianic Plague, like the Antonine Plague and Black Death, initially drove up wages. We have no numbers from Constantinople, but the literary sources are telling. In a remarkably unchristian mood, Bishop John of Ephesus complained in 544 that so many laundresses had died that the cost of getting his washing done had become outrageous (Justinian stepped in with a law pegging wages at pre-plague levels). The

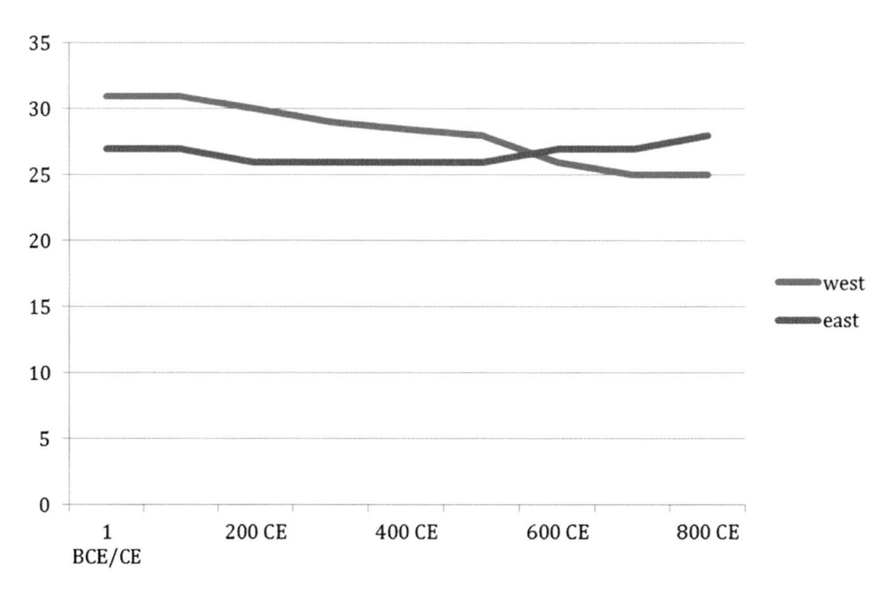

FIGURE 6.5 Eastern and Western social development scores, 1–800 (data from Morris 2013).

Middle Eastern serial data only start in the eighth century (see Figure 6.4), but several economic historians have now observed that wages at Cairo and Baghdad at that point were high, as we might expect in the aftermath of a terrible plague.[43] After 750, as the disease weakened, population recovered and wages fell for five hundred years.

Comparing the Justinianic Plague and Black Death, Pamuk and Shatzmiller conclude that 'The environment . . . in the aftermath of these plagues could stimulate increases in productivity in agriculture, urban economy, and long-distance trade by creating demand.'[44] Adding the Antonine Plague to the mix, though, highlights the differences in outcomes as well as the similarities. All three plagues brought on a combination of single-century socioeconomic collapse and rising real wages, but whereas development scores kept falling for several more centuries after the Antonine Plague (especially in the West), they began rising sharply within a century of the Black Death (again, especially in the West). The long-term effects of the Justinianic Plague in the West lie somewhere in between: after a century of rising wages and falling development, wages declined for five hundred years while development scores increased, albeit glacially slowly. By 1200 social development stood only 10 per cent above the low point reached in 700, and still languished more than 30 per cent below their first-century zenith (Figure 6.6).

The Justinianic Plague probably evolved in East Africa's Great Lakes region and spread to the Mediterranean via the Red Sea, but it had little if any impact on East Asia. There were certainly people moving back and forth in the sixth century – the mitochondrial DNA of a man named Yu Hong, buried in northwest China in 592, suggests that he was European[45] – and Chinese sources mention two outbreaks of disease in the 610s that sound alarmingly like bubonic plague, one coming across the steppes and the second by sea.[46] We do not know why the Black Death and Antonine Plague devastated both Eastern and Western Eurasia while the Justinianic ravaged only the West, but the consequence seems clear enough. While Western social development stagnated after 700, the East's kept on rising, standing 25 per cent higher in the twelfth century than it had been in the first.

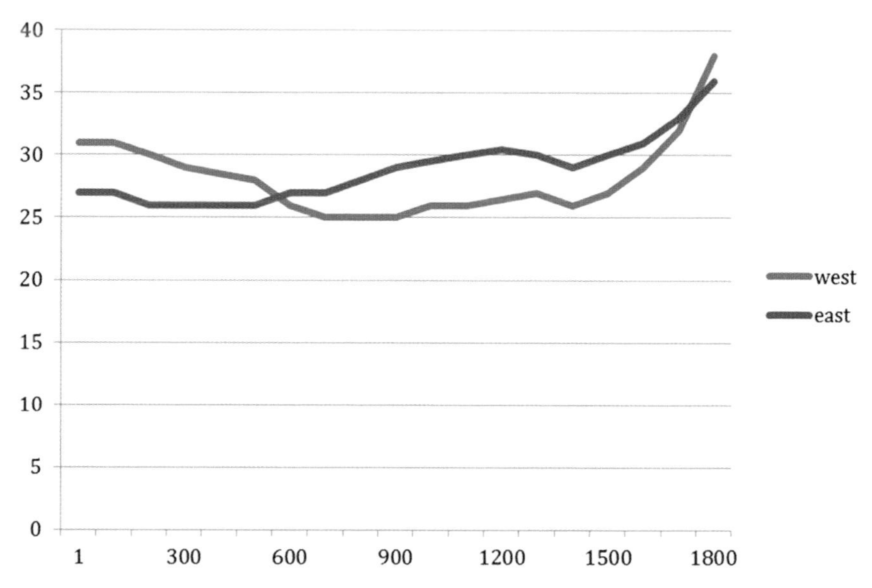

FIGURE 6.6 Eastern and Western social development scores, 1–1800 (data from Morris 2013).

There are other plagues we might compare with those of the second, sixth and fourteenth centuries. The earliest known ravaged the East Mediterranean in the fourteenth century BC. It apparently began in Egypt, whence prisoners of war carried it around 1340 BC to the Hittite Empire (in what is now central Turkey). 'The Land of Hatti, all of it, is dying', one Hittite text laments, and by the 1320s BC the sickness had reached Cyprus too.[47] This, though, is pretty much the sum total of our knowledge of the epidemic. How many died, of what causes and whether the epidemic contributed to the collapse of East Mediterranean societies a century later remain mysterious.[48]

We know much more about the plague that broke out in Athens in 430 BC. The historian Thucydides, who lived through it, described it in some detail,[49] and DNA from dental pulp suggests that it was a kind of typhoid fever.[50] Estimates vary, but it may have killed as much as one quarter of Athens' population. Diodorus of Sicily, who lived in the first century BC and was not always the most reliable of historians (Macaulay called him a 'stupid, credulous, prosing old ass'), says that plagues also ravaged the camps of Carthaginian armies besieging Syracuse in 405 and 396 BC.[51] These outbreaks, if Diodorus got the story right, may be connected to the earlier epidemic at Athens; but there is no other evidence that the sickness affected the larger Greek world. Nor is it obvious that the attested outbreaks caused collapse. Athens lost the Peloponnesian War in 404 BC and Sicily experienced a broad collapse in the 340s BC, but in most ways the fourth and third centuries BC were Greece's socioeconomic highpoint. Indeed, Greece would not see comparable levels of development again until the early twentieth century AD.[52]

The most appalling comparison case, though, is surely the 'Columbian exchange',[53] a transatlantic microbial transfer beginning in 1492 and involving everything from measles and meningitis to smallpox and typhoid. Some archaeologists think that the Native American population shrank by 90 per cent over the next two hundred years,[54] but even if the reality was closer to 50 per cent (as suggested by DNA), it is easy to see why a sixteenth-century observer concluded that 'God wishes that [the natives] yield their place to new people'.[55] By 1700, the New World had suffered one of the worst socioeconomic collapses in history – and yet in some parts of America, above all its northeast seacoast, social development was rising faster than ever. This paradox is the key to explaining whether – and when – plagues cause socioeconomic collapse.

Do Plagues Cause Socioeconomic Collapse?

I could go on adding cases, but the basic picture is probably clear. Plagues have had very varied effects. Massive excess mortality always poses a shock to a socioeconomic system, but how the system responds to that shock depends on the details of the particular case. That said, comparison of the cases also points to some larger patterns.

Many such comparisons are possible, but I will focus on just the Black Death and Antonine Plague, since they bring out the main patterns. The most striking socioeconomic consequence of the Black Death, as I noted earlier, was that a single-century collapse was followed by a multi-century recovery in both Eastern and Western Eurasia (but above all along Europe's Western shores). Labour markets responded to mass mortality much as Malthus said they should, driving up wages all across Europe and probably in the Middle East too (in China we still lack evidence) between 1350 and 1450/1500, and then driving them back down again as population recovered after 1450/1500. As wages started falling, social and political structures began diverging, with the restoration of a great empire in China, the emergence of capitalism in Europe west of the Elbe, and a second feudalism east of the Elbe. Controversies continue, but Brenner was probably largely right that local class structures explain much of the variation.

However, the East/West distinction was not the only regional contrast to develop in Europe in this period, and the data gathered by economic historians such as Robert Allen and Sevket Pamuk show that northwest European wages were diverging sharply from those across the rest of the continent by 1550/1600 (Figure 6.3).[56] In my book *Why the West Rules – for Now*, I built on their arguments to suggest that the main reason for this was the rise of a new Atlantic economy that was able to create wealth even faster than the population could grow. Throughout history, the Atlantic Ocean had functioned as a barrier cutting the Old World off from the New and isolating Western Europe from the real centres of the global economy, which lay in the band of latitudes running from the Mediterranean Sea to China. This was still the case when the Black Death broke out. Initially, between 1450 and 1500, Northwest European wages did start falling in response to population growth; but between 1550 and 1600, the Atlantic began to change its meaning, with massive consequences. New kinds of ships changed the Atlantic from a barrier isolating Northwest Europe into a highway connecting it to the Indian Ocean and the Americas. An entirely new kind of market economy began to boom, centred on England and the Netherlands, the countries best placed to exploit the Atlantic highway; and in these countries wages began rising again.[57]

The most striking socioeconomic consequence of the Antonine Plague, by contrast, was a multi-century collapse, lasting until about 450 in China and 750 in the Mediterranean basin. As in the case of the Black Death, labour markets probably responded in broadly Malthusian ways, with wages rising in the late second century and falling in the third. Once again, though, the changing meanings of geography shaped the plague's larger consequences. In China, war and conquest began linking the booming rice production centres south of the Yangzi River with the great cities of the Yellow River plains in the fifth and sixth centuries. The reunification of China into a single empire in 589 accelerated economic integration, and in the seventh century the Grand Canal sharply reduced the costs of shipping rice to the North. The boom that these centuries saw was smaller than that enjoyed by Northwest Europe after 1600, but it produced similar results: because the economy grew faster than population, socioeconomic collapse was averted.[58]

In Western Eurasia, geography also changed its meaning in these centuries, but it moved in the opposite direction. For many centuries before the Antonine Plague, the Mediterranean Sea had been the busiest trade zone on earth, driving the Roman Empire's social development up to the highest levels yet seen. After the 160s, however, this framework began unravelling. In the fifth century, Germanic invasions broke the trade links between Italy and North Africa, and in the seventh century

Persian and then Arab invasions did the same to the link between Constantinople and Egypt. Back in the 1930s, the medievalist Henri Pirenne suggested that the Arab conquests marked the real end of the ancient world, sundering the economic ties that had maintained Mediterranean prosperity.[59] Since the 1980s, archaeologists and historians have regularly insisted that Pirenne was wrong, but even the critics' own accounts make it clear that the quantity of trade, the size of exchange networks and cities, the use of coins and standards of living all dropped sharply (if unevenly) between 200 and 700.[60] The Mediterranean Sea changed its meaning, from being a highway uniting the lands around it to being a barrier dividing them. The economy did not grow faster than the population in the centuries after the Antonine (and Justinianic) Plague and socioeconomic collapse ensued.

The other plagues mentioned in the previous section all seem to underline the same lesson. Plagues shock societies by killing huge numbers of people. When the shock is big enough, institutions break down and social development scores fall, although mass mortality simultaneously shifts the land: labour ratio in favour of workers, driving up real wages. The wage gains are normally a short-term outcome, because within three generations, population starts growing again. The land: labour ratio then shifts back, and real wages fall. When that happens, socioeconomic collapse may continue for centuries, especially if (as happened with the Justinianic Plague in the sixth century) the system suffers further shocks.

The great exception to this pattern is Northwest Europe after 1600, where the conversion of the Atlantic into a highway meant that the economy grew faster than the population. Real wages kept moving upwards and socioeconomic collapse turned into a powerful recovery. Although this was centred on Northwest Europe, Northwest European colonists in North America also benefitted hugely, even while Native Americans were suffering a multi-century socioeconomic collapse; and the post-Black Death revival in East Asia was at least partly driven by connections to the new Atlantic economy.[62] Something rather similar (but on a smaller scale) probably explains China's escape from socioeconomic collapse after 500, and perhaps also the Greek experience in the fourth century BC.

Plague and Socioeconomic Collapse in the Twenty-First Century

There are many conclusions to be drawn from the long-term history of plagues. I will begin with some bad news: the story seems to imply that global pandemics are likelier in the twenty-first century than ever before.[63] In the past, human mobility has been the mother of plague – the more that people move around, the more they mix previously separate disease pools, exposing epidemiologically virgin populations to new pathogens. Since the industrial revolution, mobility and the spread of disease have accelerated together. The steamships that carried millions of young men across oceans spread the H1N1 virus around the world in 1918–19. AIDS, after remaining isolated in Central Africa from at least 1959 until the early 1980s, abruptly leapt to four continents with the help of jet travel. In 2003, Severe Acute Respiratory Syndrome (SARS) reached thirty-seven countries within weeks of evolving in southern China, and in 2009 the so-called New H1N1 had arrived on four continents before it was even recognised.[64]

The good news, though, is that the same soaring social development that gave us the means to speed up the merging of disease pools has also given us the medical skills to respond quickly and the organisational tools to fight back. In 1918–19, there was nothing to do but let H1N1 run its course. When AIDS broke out from Africa in the late twentieth century, by contrast, medical researchers could sequence the genome of HIV (human immunodeficiency virus) that caused it, even if this took fifteen years; and in 2003, SARS was sequenced in just fifteen days, and aggressive international action nipped a potential plague in the bud.

New strains of influenza – H7N9, H10N8[65] – are constantly evolving. The World Health Organization (WHO) estimates that if any of these were to break out and then act like the H2N2 virus (which killed 1–2 million people in 1957), it would mean somewhere between 2 million and 7.4 million deaths; if it behaved more like the 1918 H1N1, it could kill up to 200 million. In 2005, the WHO published a list of 'Ten things you need to know about pandemic influenza', which included the following terrifying facts:

> The world may be on the brink of another pandemic
> All countries will be affected
> Medical supplies will be inadequate
> Large numbers of deaths will occur
> Economic and social disruption will be great.[66]

No one knows whether we will keep winning the race between medicine and microbes,[67] or whether – if we lose the race – a twenty-first-century plague would set off a multi-century socioeconomic collapse like those that followed the Antonine and Justinianic plagues. Long-term history, however, can perhaps limit the uncertainty.

I suggested earlier that the main reason that the Black Death produced a single-century collapse followed by a strong recovery while the Antonine Plague set off a multi-century collapse followed by a weak recovery was that when population started growing again after the Black Death, the economy grew even faster (especially in northwest Europe), whereas when population revived after the Antonine Plague, the economy did not grow faster. In some ways, it seems to me that a new plague is more likely to have Black Death-type effects than Antonine Plague-type effects. If hundreds of millions of people do die from disease and starvation in the immediate aftermath of a pandemic, institutions will surely fail, social development will slump and technological skills will be lost. A collapse lasting a century seems perfectly possible. However, even if a new plague were to cut the planet's population and social development by 10–20 per cent over the next hundred years – huge figures, even by the standards of the Black Death – the survivors would still live in a world of 5.5 billion people with a social development score well above 700 points. There would still be engineers, who would still understand fossil fuels, nuclear power and electricity. Other things being equal, I suspect that a twenty-first-century plague and socioeconomic collapse would lead to a massive twenty-second-century recovery.

The problem, though, is that history also suggests that other things will not be equal. In the past, socioeconomic collapses have always involved the same five proximate causes. In my book *Why the West Rules – For Now*, I labelled these 'the five horsemen of the apocalypse'.[68] First was mass migration, beyond the capacities of the governments of the day to manage. Second was epidemic disease, often caused by the merging of previous separate disease pools. Third was state failure and war, as the shocks of migration and plague brought governments down. Fourth was famine, as trade routes dried up in the face of rising violence; and fifth was climate change, which always stood in a complicated relationship to the other four horsemen, and yet was always present when disasters crossed the line into collapse.

It is difficult these days even to glance at a news magazine or website without getting the feeling that all five horsemen are once again closing in, and that we are about to rerun the familiar script of collapse (whether of the single- or the multiple-century variety). However, as Mark Twain is supposed to have observed, history does not repeat itself; it merely rhymes.[69] Every earlier collapse has involved state failure and mass violence, but every time humanity has rerun the plague-and-collapse narrative it has done so in a new way. When states failed in the third, seventh and fourteenth centuries, iron-armed infantry defended fortified cities against battering rams and bows and arrows; but when twenty-first-century states go to war, they can use nuclear and biological weapons.

In 1983, the US Strategic Concepts Development Center found out just how bad that would be when it ran a series of war games simulating the opening rounds of a nuclear exchange with the Soviet Union. The Center instructed the players on both sides to follow current nuclear doctrines, which meant making counterforce (i.e., firing just at enemy military targets, not population centres) rather than countervalue (i.e., firing on cities) attacks. When it came to it, though, no player managed to draw the line at counterforce. Every time the game was run, everyone involved escalated to countervalue strikes. The algorithms behind the game implied that the death toll would quickly climb to half a billion, and that fallout, starvation and further fighting would kill another half billion in the weeks and months that followed. Beyond that point, no one could say what might happen, but extinction was a real possibility.[70]

Weapons of mass destruction threaten to yield a whole new kind of socioeconomic collapse, not just multi-century but permanent. That said, the fact we have the power to destroy ourselves does not necessarily mean that we will, and some long-term trends seem to point exactly the opposite way. Ethnographic and archaeological evidence suggests that rates of violent death have declined by an order of magnitude across the last ten thousand years. The average person in the Stone Age stood a 10–20 per cent chance of dying violently, but in the twentieth century – despite its world wars, nuclear attacks and genocides – that figure had fallen to just 1–2 per cent. According to the WHO, it now stands at just 0.7 per cent.[71] We live in a kinder, gentler world than ever before, in which people are much less likely to seek violent solutions to their problems.

Whether this means that next time really will be different, and that state failure will not generate war in a twenty-first century socioeconomic collapse, of course depends on the causes of the decline in violence. In my recent book *War: What Is It Good for?* I suggested that despite the messiness of the details of the decline, the underlying explanation is actually very simple.[72] Basically, violence is an evolved but suboptimal adaptation shared by most species of animals; and among humans, the only species capable of cultural as well as biological evolution, violence has been putting itself out of business for the last ten thousand years.

What I mean by this rather cryptic formulation is that in the ten millennia since agriculture was first established, groups that won wars regularly swallowed up those that lost, forming larger societies. The men who ruled these larger societies found that competition (both with other states and with rivals within their own states) put pressure on them to pacify their realms – not because ancient kings were angels, but because rulers with law-abiding, tax-paying subjects generally fared better than rulers with murderous, rebellious ones. Not all wars advanced the trend towards larger and safer societies, of course, but enough did that rates of violent death moved slowly (and unevenly) downwards across ten millennia. War made the state, and, as Hobbes saw in the seventeenth century, the state made peace.[73] More to the point – as Hobbes at least partly recognised – war is the only force that has consistently made states, and states are the only force that has consistently made peace.[74] If this argument is correct, it means not only that the long-term trend away from killing can be reversed,[75] but also that state failure – which has always accompanied socioeconomic collapse in the past – is precisely the force likely to effect a reversal.

Once again, there is bad news and good news. The bad news is that if the twenty-first century does see plague and socioeconomic collapse, it is almost certain to see more wars too, quite possibly involving nuclear weapons. The good news comes in two parts. The first is that the worst-case scenario is nowhere near as bad as it used to be. If plague and socioeconomic collapse were to set in tomorrow (I am writing in summer 2014) and state failure brought on nuclear war, even an all-out exchange could only kill a fraction of the billion-plus who would have died in the 1980s. For every twenty nuclear warheads in the world back then, there is now only one.[76] Nor can the great powers easily increase their nuclear arsenals. In 2013, the United States even put its new Los Alamos plutonium production facility on hold because of budget problems, ruling out any rapid rearmament.[77] We can no longer kill everyone at once, and even if the worst comes to the worst, we might get away with a multi-century or even a single-century collapse rather than a permanent one.

The second part of the good news is that soaring social development is changing the world in ways that even a serious socioeconomic collapse might be unable to undo. By this I mean not so much the functional integration that globalisation has brought, because this, if earlier collapses are anything to judge by, will be among the first things to go; rather, I have in mind the technological transformation that is beginning to link people around the world in ways that were unimaginable even fifty years ago. People are merging with their machines, and, through their machines, with each other.[78] In 2012, neuroscientists linked the brain of a rat in Brazil via the Internet to that of a rat three thousand miles away in North Carolina, allowing the first rat to move the second rat's paws (70 per cent of the time, at least). In 2014, a computer in Japan managed to simulate human brain activity (although the experiment modelled just 1 per cent of the brain's neural network, and needed forty minutes to do the calculations that a brain does in a single second). 'There's a long way to go before you get to proper mind-reading', says Jan Schnupp, who holds the chair in neuroscience in Oxford University; but 'it's a question of when rather than if . . . It is conceivable that in the next ten years this could happen.'[79]

Whether humanity is now beginning to meld into a single super-organism, as the science writer Robert Wright predicts, or whether such talk is just 'digito-futuristic nonsense' and a 'Cyber-Whig theory of history' (as the influential technology critic Evgeny Morozov puts it), is of course open to debate,[80] and whatever happens next, the technological merging of mankind will surely bring costs of its own. In fact, the linkage of biological life and technology may constitute the next mixture of disease pools leading to new epidemics. This may be either through the implementation of electronic devices into biological organisms, through interfaces between brains and computers, or more profoundly through the machines producing biological material and in particular DNA. That said, it seems likely to me that the further we move in the Schnupp-Wright direction, the less chance there will be that the shock of plague and socioeconomic collapse on a one-century scale will generate nuclear reactions that lead to socioeconomic collapse on the permanent scale.[81] But only time will tell who is right.

Conclusions

So, to conclude, I return to my opening question: do plagues cause socioeconomic collapse? My answer: yes and no. They act as shocks, sometimes massive, to socioeconomic systems. In my survey of the last fifteen thousand years of history I found no definite examples of plagues that crossed the 10 per cent/5 per cent threshold described earlier without bringing on at least a single-century collapse (in the sense of a decline in social development of at least 10 per cent), and no definite examples of 10 per cent declines in development without plagues being involved.[82] In this sense, we can say that plagues are both necessary and sufficient causes of single-century socioeconomic collapses. Where evidence survives, though, single-century collapses were often good for ordinary people's well-being, shifting the land:labour ratio in favour of workers.

Multi-century collapses are more complicated. I know of no example that did not involve a plague, so plagues do seem to be a necessary condition; but the story of Eurasia since the Black Death shows that plagues are not a sufficient condition. The decisive factor in turning a single-century collapse into a multi-century one seems to be the growth of the economy. When the economy grows faster than the population recovers after the mass mortality, socioeconomic bust turns to boom; when it does not, single-century collapse turns into the multi-century version.

Long-term history suggests that new plagues are perfectly possible in the twenty-first century, and our modern mastery of the means of destruction gives us the option of turning even a small collapse into a permanent one. But all things considered, that does not seem very likely, and while there is probably no way to end a chapter about plague and socioeconomic collapse on an upbeat note, we can at least console ourselves with one final piece of good news: things could be worse.

References

1. Ibn Khaldûn, *Muqadimmah* 1.64, cited from Dols Michael. (1976) *The Black Death in the Middle East.* Princeton: Princeton University Press. All dates in this essay are AD unless marked BC.
2. Benedictow Ole (2004) *The Black Death 1346–1353: The Complete History.* Rochester, NY:Boydell Press; Dols (1976); McNeill William (1976) *Plagues and Peoples.* New York:Anchor:132–75.
3. Villani Matteo *Description of the Plague in Florence* (1348), quoted from Kirschner Julius, Karl Morrison, eds. (1986). *University of Chicago Readings in Western Civilization IV: Medieval Europe.* Chicago: University of Chicago Press:448.
4. Population: Biraben J.R. (1979). Essai sur l'évolution du nombre des hommes. *Population* 34:13–25.; Longevity:Livi-Bacci Massimo (2001) *A concise history of world population.* 3rd edn. Trs. Carl Ipsen. Oxford: Wiley-Blackwell.; Wealth: Maddison Angus (2010) Statistics on world population, GDP, and per capita GDP, 1–2008 AD. www.ggdc.net/maddison/Maddison.htm. Accessed 15 May 2015.
5. I would like to take this opportunity to thank once again Mary Fowler, Jonathan Heeney, Sven Friedemann and Janet Gibson for inviting me to give a Darwin Lecture and for making my time at the college in February 2014 so enjoyable, and to Anthony Snodgrass, Annemarie Künzl, Paul Cartledge and Peter Garnsey for their conversation. I should add that the organisers originally invited me to speak about 'Plagues and economic collapse', but on reflection, the less elegant 'Plagues and socioeconomic collapse' seems a better fit with what I actually said.

6. Published as Gellner Ernest (2005) Origins of society. In A.C. Fabian, ed., *Origins*: 128–40. Cambridge: Cambridge University Press.
7. Morris Ian (2015) *Foragers, Farmers, and Fossil Fuels: How Human Values Evolve*. Princeton: Princeton University Press.
8. In Michael Postan's classic book *The Medieval Economy and Society*, for instance, under 'Black Death' the index simply says '*see* Plague' (Postan Michael (1972) *The Medieval Economy and Society: An Economic History of Britain in the Middle Ages*. London:Weidenfeld & Nicolson).
9. Tainter Joseph (1988) *The Collapse of Complex Societies*. Cambridge: Cambridge University Press: 4 (emphasis in original); Diamond Jared (2005) *Collapse: How Societies Choose to Fail or Succeed*. New York: Viking:3.
10. Morris Ian (2010) *Why the West Rules – For Now: the Patterns of History and What They Reveal about the Future*. London:Profile; (2013) *The Measure of Civilisation: How Social Development Explains the Fate of Nations*. London:Profile.
11. Morris (2010):144; 2013:5.
12. Available at http://hdr.undp.org/en.
13. ul Haq 1995.
14. Index making is, admittedly, chainsaw art rather than Old Master material, and the margins of error may well run as high as 10 per cent (although we can be fairly sure that they do not run much higher [Morris 2013:239–52]). Every step of the process also raises major definitional problems, which I examine at Morris (2013):17–52.
15. Morris (2010):393–410. As always in historical analysis, however, multiple factors were interconnected. Eastern development was already falling in the thirteenth century, largely because of Jurchen and Mongol invasions of China from the steppes; but since steppe migrations were the main mechanism for spreading the Black Death after 1331, we cannot really separate migration and disease. Migrants began the collapse by devastating agrarian societies, and then accelerated it by merging disease pools across Eurasia.
16. The historical literature is huge and full of subtle distinctions, but the core argument goes back to Postan Michael (1950) Moyen âge. In *Rapports du IXe congrès international des sciences historiques I*: 225–41. Paris:A. Colin, with important qualifications and elaborations in Habakkuk H. J. (1958) The economic history of modern Britain. *Journal of Economic History* 18:486–501, Postan Michael (1966) England. In Michael Postan, ed., *The Cambridge Economic History of Europe I: Agrarian Life of the Middle Ages*:548–632. Cambridge: Cambridge University Press, and Le Roy Ladurie Emmanuel (1966)

Les paysans de Languedoc. 2 vols. Paris:SEVPEN, and a more up-to-date version in Herlihy David (1997) *The Black Death and the Transformation of the West*. Cambridge, MA: Harvard University Press.

17. Particularly Dobb Maurice (1946) *Studies in the Development of Capitalism*. London:Routledge; Hilton Rodney, ed. (1976) *The Transition from Feudalism to Capitalism*. London:Verso; Bois Guy (1984 [1976]) *The Crisis of Feudalism*. Cambridge: Cambridge University Press.
18. Brenner Robert (1985 [1976]). Agrarian class structure and economic development in pre-industrial Europe. In Aston and Philpin 1985:10–63. First published in *Past and Present* 70:30–74::21, 23.
19. Kula Wittold (1974) *An Economic Theory of Feudalism*. London:Verso.
20. Particularly the papers in Aston T.H., C.H.E. Philpin, eds. (1985) *The Brenner Debate: Agrarian Class Structure and Economic Development in Pre-Industrial Europe*. Cambridge: Cambridge University Press.
21. Allen Robert (2001) The great divergence in European wages and prices from the Middle Ages to the First World War. *Explorations in Economic History* 38:411–48; Clark Gregory (2005) The condition of the working class in England, 1209–2008. *Journal of Political Economy* 113:1307–40; Pamuk Sevket (2007) The Black Death and the origins of the 'great divergence' across Europe, 1300–1600. *European Review of Economic History* 11:298–317. Scheidel Walter (2010) Real wages in early economies: evidence for living standards from 1800 BCE to 1300 CE. *Journal of the Economic and Social History of the Orient* 53:425–62 identifies data going back to 1800 BC, but no continuous series runs from ancient to modern times.
22. Heinrich Müller, cited in Braudel Fernand (1981) *Civilization and Capitalism, 15th–18th Century I: The Structures of Everyday Life*. Trs. Siân Reynolds. New York:Harper and Row.
23. The papers in Allen Robert et al., eds. (2005) *Living Standards in the Past: New Perspectives on Well-Being in Asia and Europe*. Oxford:Oxford University Press and Bengtsson Tommy et al., eds. (2005) *Life Under Pressure: Mortality and Living Standards in Europe and Asia, 1500–1700*. Cambridge: Cambridge University Press collect much material.
24. Pamuk Sevket, Maya Shatzmiller (2014) Plagues, wages, and economic change in the Islamic Middle East, 700–1500. *Journal of Economic History* 74:196–229.
25. Allen Robert et al. (2011) Wages, prices and living standards in China, 1738–1925: A comparison with Europe, Japan and India. *Economic History Review* 64 Supplement: 8–38 tell the Chinese story only from 1738 onwards.

26. Zhang Tao, *Gazetteer of She County* 6.10b–12a, cited from Brook Timothy (1998) *The Confusions of Pleasure: Commerce and Culture in Ming China*. Berkeley:University of California Press.:1, 4.
27. McNeill 1976.
28. McNeill 1976:5.
29. MacKellar Landis (2007) Pandemic influenza: A review. *Population and Development Review* 33:429–51.
30. www.unaids.org/en/resources/campaigns/globalreport2013/factsheet/
31. Cited from McNeill 1976, p. 118.
32. The details are debated by the contributors to Lo Cascio Elio, ed. (2012) *L'impatto della 'peste antonina'*. Bari:Edipuglia. In the earlier scholarship, Duncan-Jones 1996 defends a high mortality estimate, while Greenberg Joseph (2003) Plagued by doubt: Reconsidering the impact of a mortality crisis in the second century A.D. *Journal of Roman Archaeology* 16:413–25 and Bruun 2003 and 2007 prefer lower estimates.
33. Tiradritti Francesco (2014) Of kilns and corpses: Theban plague victims. *Egyptian Archaeology* 44
34. Scheidel Walter (2002) A model of demographic and economic change in Roman Egypt after the Antonine Plague. *Journal of Roman Archaeology* 15:97–114; (2012) Roman wellbeing and the consequences of the Antonine Plague. In Lo Cascio 2012: 265–95; Sallares 2007.
35. McNeill 1976:96–109.
36. DNA:www.independent.co.uk/news/science/archaeology/news/ambassador-or-slave-east-asian-skeleton-discovered-in-vagnari-roman-cemetery-1879551.html. Texts: Leslie D.D., K.J.H. Gardiner (1996) *The Roman Empire in Chinese Sources*. Rome:Bardi.
37. The first and second centuries AD, like the thirteenth and fourteenth, also saw a great increase in maritime contact between the Middle East and India, but the Antonine Plague seems to have had no more impact on South Asia than the Black Death would do. This suggests that in both periods, horses on the steppes, rather than ships on the Indian Ocean, were the main vectors for merging eastern and western disease pools.
38. I explain my own views on the Antonine Plague in Morris 2010:292–8.
39. Scheidel 2012:288.
40. In a comparative study of the Black Death in England and Egypt, Borsch Stuart (2005) *The Black Death in Egypt and England:*

A Comparative Study. Austin: University of Texas Press suggests that Egypt after 1350 in fact began a long-term socioeconomic decline very similar to the one I suggest came in the wake of the Antonine Plague.

41. Harbeck Michaela et al. (2013) Yersinia pestis DNA from skeletal remains from the 6th century reveals insights into Justinianic Plague. *PLoS Pathogens* 9(5):e1003349; Wagner David et al. (2014) Yersinia pestis and the Plague of Justinian 541–543 AD: A genomic analysis. *The Lancet Infectious Diseases* 14(4):319–26.
42. Keys David (2000) *Catastrophe: An Investigation into the Origins of Modern Civilization*. New York:Ballantine; Little Lester, ed. (2007) *Plague and the End of Antiquity: The Pandemic of 541–750*. Cambridge: Cambridge University Press; Rosen Stanley (2007) *Justinian's Flea: Plague, Empire, and the Birth of Europe*. New York:Viking; Sarris Peter (2002) The Justinianic Plague: Origins and effects. *Continuity and Change* 17:169–82; Stathakopoulos Dionysios (2007) *Famine and Pestilence in the Late Roman and Early Byzantine Empires*. Burlington, VT:Ashgate.
43. Ashtor Eliyahu (1969) *Histoire des prix et des salaires dans l'orient médiéval*. Paris:SEVPEN; Findlay Ronald, Mats Lundahl (2006) Demographic shocks and the factor proportion model: from the Plague of Justinian to the Black Death. In Ronald Findlay et al., eds., *Eli Heckscher, International Trade, and Economic History*:157–98. Cambridge, MA: Harvard University Press; Scheidel 2010; Pamuk and Shatzmiller 2014.
44. Pamuk and Shatzmiller 2014:221–2.
45. Xie C.Z. et al. (2007) Evidence of ancient DNA reveals the first European lineage in Iron Age central China. *Proceedings of the Royal Society B* 274:1597–1601.
46. Twitchett Denis (1979) Population and pestilence in T'ang China. In Wolfgang Bauer, ed., *Studia Sino-Mongolica: Festschrift für Herbert Franke*: 35–68. Wiesbaden:Harrassowitz Verlag.
47. Bryce Trevor (1998) *The Kingdom of the Hittites*. Oxford: Oxford University Press:223–5; Cline 2011:60.
48. Cline Eric (2014) *1177 B.C.: The Year Civilization Collapsed*. Princeton: Princeton University Press has a good review of the collapse, which cut Western social development by 8 per cent between 1300 and 1000 BC.
49. Thucydides 2.47–54. Thucydides (3.27) also mentions a recurrence in 427/6 BC.
50. Papagrigorakis Manolis et al. (2006) DNA examination of ancient dental pulp incriminates typhoid fever as a probable cause of the plague of Athens. *International Journal of Infectious Diseases* 10:206–14.

51. Diodorus 13.114; 14.63, 70–1. Macaulay is quoted in Green Peter (2010) *Diodorus Siculus*, books 11–12.37.1. Austin: University of Texas Press:ix., but Green defends Diodorus as 'a rational, methodical, if somewhat unimaginative, minor historian' against what he calls the 'near-hysterical academic chorus of dismissal' (2010:x).
52. Ober Josiah (2015) *The Rise and Fall of Classical Greece*. Princeton: Princeton University Press.
53. A term coined in Crosby Alfred (1972) *The Columbian Exchange: Biological and Cultural Consequences of 1492*. Westport, CT:Westview Press; and, (2003) Ecological Imperialism: The Biological Expansion of Europe, 900–1900. 2nd edn. Cambridge:Cambridge University Press.
54. Mann Charles (2005) *1491: New Revelations of the Americas before Columbus*. New York:Knopf.
55. DNA: O'Fallon Brendan, Lars Fehren-Schmitz (2011) Native Americans experienced a strong population bottleneck coincident with European contact. *Proceedings of the National Academy of Sciences* 108:20444–8. Quotation from Crosby 2003:215.
56. Allen 2001, 2009; Pamuk 2007.
57. Morris 2010:459–68.
58. Morris 2010:332–42.
59. Pirenne Henri (1939 [1937]) *Mohammed and Charlemagne*. Trs. Bernard Miall. New York:Norton.
60. E.g., Hodges Richard, David Whitehouse (1983) *Mohammed, Charlemagne, and the Origins of Europe: Archaeology and the Pirenne Thesis*. London:Duckworth; McCormick Michael (2001) *Origins of the European Economy: Communications and Commerce, AD 300–900*. Cambridge: Cambridge University Press; Wickham Chris (2005) *Framing the Early Middle Ages: Europe and the Mediterranean, 400-800*. Oxford: Oxford University Press; 2009 *The inheritance of Rome: illuminating the Dark Ages, 400–1000*. New York:Penguin; Abulafia David (2011) *The Great Sea: A Human History of the Mediterranean*. London:Allen Lane.
61. Morris 2010:356–63.
62. I expand on this theme in Morris 2010:449–55.
63. Garrett Laurie (1994) *The Coming Plague: Newly Emerging Diseases in a World out of Balance*. New York:Farrar, Straus & Giroux remains a valuable study.
64. New H1N1: www.who.int/influenza/resources/publications/evolution_pandemic_Ah1n1/en/.
65. www.who.int/influenza/human_animal_interface/influenza_h7n9/en/; www.thelancet.com/journals/lancet/article/PIIS0140-6736(14)60163-X/fulltext.

66. www.who.int/wer/2005/wer8049.pdf?ua=1.
67. See www.washingtonpost.com/national/health-science/us-launches-new-global-initiative-to-prevent-infectious-disease-threats/2014/02/12/afd9863c-936d-11e3-b46a-5a3d0d2130da_story.html; www.who.int/influenza_vaccines_plan/resources/progress_materials/en/, and the forum on 'Winning the battle against emerging pathogens' in *Bulletin of the Atomic Scientists* 70.4 (2014), available at http://thebulletin.org.
68. Morris 2010:223–6.
69. No one has ever tracked Twain's supposed comment ('History doesn't repeat itself, but it does rhyme') back to a source, so I have felt free to adapt it to fit my sentence.
70. Bracken Paul (2012) *The Second Nuclear Age: Strategy, Danger, and the New Power Politics.* New York:Time Books.
71. Pinker Steve (2011) *The Better Angels of our Nature: Why Violence Has Declined.* New York:Viking; Morris Ian (2014) *War: What Is It Good for? Conflict and the Progress of Civilisation from Primates to Robots.* London:Profile; www.who.int/violence_injury_prevention/violence/en/.
72. I emphasise this (Morris 2014:319–25) in distinction to the more complicated 'tale of six trends, five inner demons, and five historical forces' offered by Steven Pinker (2011:xxiv).
73. This argument inverts Charles Tilly's famous claim that 'war made the state, and the state made war' (Tilly Charles (1975) Reflections on the history of European state-making. In Charles Tilly, ed., *The Formation of National States in Western Europe*:3–83. Princeton: Princeton University Press).
74. Every sentence (perhaps every word) in this paragraph is hotly contested. I explain my views in detail in Morris 2014.
75. There is even a good historical analogy for such a reversal; for 1200 years, between roughly AD 200 and 1400, Eurasian wars went through a long period in which instead of combining small violent societies into large peaceful ones, they generally did the opposite (Morris 2014:112–64).
76. Kristensen Hans, Robert Norris (2013) Global nuclear weapons inventories, 1945-2013. *Bulletin of the Atomic Scientists* 69(5), available at http://thebulletin.org/2013/september/global-nuclear-weapons-inventories-1945–2013; (2014a) US nuclear forces, 2014. *Bulletin of the Atomic Scientists* 70(1), available at http://thebulletin.org/2014/january/us-nuclear-forces-2014; (2014b) Russian nuclear forces, 2014. *Bulletin of the Atomic Scientists* 70(2), available at http://

thebulletin.org/2014/march/russian-nuclear-forces-2014. The 95 per cent decline counts only active warheads; counting older warheads that have not yet been dismantled, the figure is closer to 80 per cent.

77. www.lasg.org/press/2013/NWMM_22Feb2013.html.
78. I discuss this trend in more detail in Morris 2010:590–8.
79. Rats: www.nature.com/srep/2013/130228/srep01319/full/srep01319.html. Japanese computer: www.hpcwire.com/2014/01/15/supercomputer-models-human-brain-activity/. Schnupp.
80. Wright Robert (2000) *Nonzero: the Logic of Human Destiny*. New York:Pantheon; Morozov, at www.newrepublic.com/article/books-and-arts/magazine/105703/the-naked-and-the-ted-khanna#.
81. I expand on this argument at Morris 2014:377–91.
82. I stress 'definite examples' because there are cases – such as the disintegration of the Indus Valley civilisation after 1900 BC, the East Mediterranean breakdown after 1200 BC and the Classic Maya collapse after AD 600 – about which we just do not know enough to specify the role of disease.

7 Silicon Plagues

MIKKO HYPPONEN

Digital technology is changing at a faster pace than any other part of our world. The development of the integrated circuit started a revolution that eventually led to the development of technologies like packet-switched networks, which were required to create the internet and the worldwide web. It's easy to see how many beneficial things personal computers and the internet have brought us. They have changed not only our communication and the way we entertain ourselves, but also the way we think. Unfortunately, they have also brought us new kinds of risks.

The online world is a reflection of the real world. Just like we have crime in the world, we have crime in the online world as well. The big difference is that distances and country borders do not exist in the online world – we are not safe from an online criminal just because he's living faraway. Today, our world is largely dependent on digital networks, from personal banking to stock markets to military systems.

How real is the risk of a digital Plague?

Very real.

We've already been fighting computer viruses and other types of malware for decades. They have evolved from simple, straightforward attacks to complicated, global outbreaks. In some ways, the evolution of online attacks resembles biological evolution. But the difference is that all computer attacks are created and launched by humans. By looking at examples of key malware attacks through the years, we can see how attacks on our digital world have evolved over time.

The Origins of Computer Viruses

Brain.A[1] is considered to be the first PC virus in history. It was first detected in 1986. Several variants of the virus followed but most of them were fairly harmless. It ran on IBM-PCs and compatibles with PC-DOS operation systems. Brain was a boot sector virus, infecting the first sector of floppy discs as they were inserted into an infected computer. Brain was only a few kilobytes in size. Before Brain infected diskettes, it looked for a 'signature'. This made it possible to 'inoculate' against the virus by putting the signature in the correct place of the boot sector of a clean floppy. Such floppies would not get infected even if they were inserted into an infected computer. The Brain virus tried to hide from detection by hooking the operating system functions that were used to read the floppy drive. When an attempt was made to read an infected boot sector, Brain would show you the original boot sector instead. This meant that if you looked at the boot sector, everything looked normal, even if the virus was active in memory. The major effect of Brain.A was a change of the disk label (the 'name' of the disk). The volume label was changed to read: '©Brain'. The code of the virus contained a hidden text section, listing the address of the creators of the virus (Figure 7.1).

Brain was quickly followed by other simple boot and file sector infections, such as Stoned, Cascade, Yankee Doodle, Dark Avenger and Form. We have to remember that in the late 1980s PC computers were not connected to networks. In fact, normal PCs were not connected to each

FIGURE 7.1 Snippet from a boot sector infected by Brain.A.

other in any way. The only practical way of moving data was via floppy disks. This is why floppy-based viruses (such as Brain, Stoned and Form) spread so quickly.

Infections with viruses like Brain, Stoned and Form were invisible to the victim. The viruses would not show up in any way. When we look at some of the most common viruses worldwide during the first malware years, we see that most of them had no visible activation feature at all as summarised in Table 7.1.

While most early viruses were invisible, some of them did make themselves known to the victims (Table 7.2).

The Destructive Nature of Early Computer Viruses and How They Were Triggered

Examples of immediate destruction include viruses like Michelangelo, Kampana and Natas, which simply overwrote part of the hard drive. Other viruses with immediate destructive routines would delete or overwrite files instead of overwriting physical sectors. Gradual destruction was done by viruses such as Ripper or Nomenklatura, which slowly corrupted the data on the hard drive. Such corruption was likely to go unnoticed until the corrupted data had been backed up several times. This made recovering

Table 7.1

AntiCMOS.A	had an activation routine, which was never executed
AntiEXE	had an activation routine, which was practically never executed
DIR_II.A	no activation routine
Form.A	no activation routine
Tai-Pan.438	no activation routine
Junkie	no activation routine
Stoned.Empire.Monkey.B	no activation routine
Stoned.Standard.A	had an activation routine, which was very seldom executed
Stoned.No_INT.A	no activation routine
Stealth_Boot.B	no activation routine

Table 7.2

Kampana.A	overwrites part of the hard drive after 400 boots
Green_Caterpillar.1575 (Figure 7.2)	draws a Caterpillar on screen after 60 days
Michelangelo	overwrites part of the hard drive on every 6th of March
Cascade.1701.A	drops letters to the bottom of the screen
V-Sign	draws an ASCII graphics of a large V after every 64 boots
Tequila (Figure 7.3)	draws a fractal by random

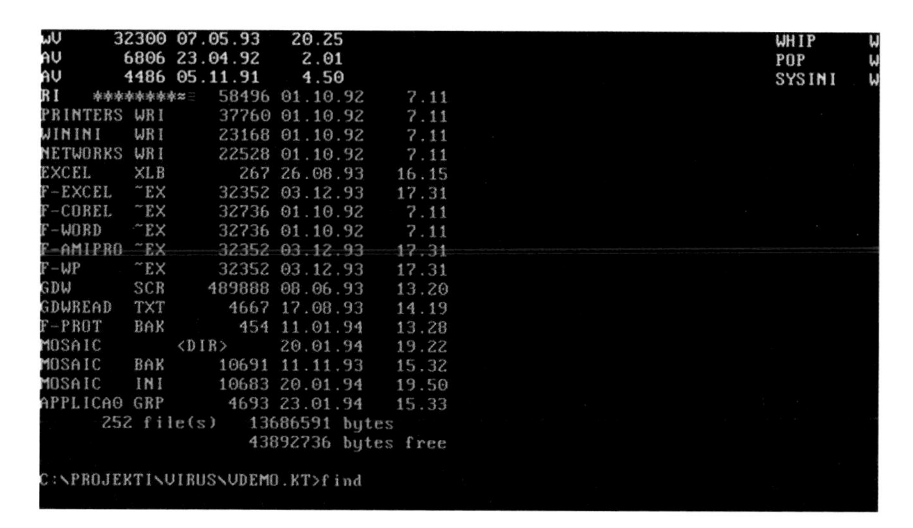

FIGURE 7.2 Activation routine of Green_Caterpillar.1575.

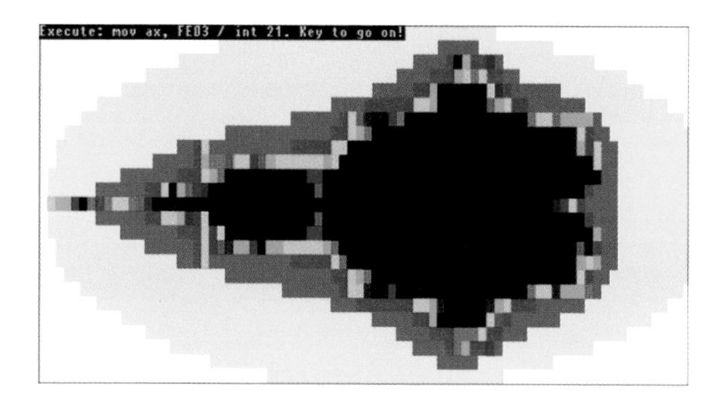

FIGURE 7.3 Fractal display of the Tequila virus.

considerably more difficult, and in most cases significant amounts of data were lost for good. Thankfully, destructive activation routines quite often failed to work due to programming errors. It seems that the virus authors were reluctant to test these routines on their own machines.

Sounds, Tunes and Speech

There were several early viruses which activated by playing tunes through the PC speaker. Probably the most common examples were the different Yankee_Doodle variants, which played the Yankee Doodle tune at different times of day. Other viruses just produced beeps occasionally. There were also some viruses that tried to speak via the PC speaker.

Animations

Viruses that activated with an animation can be further grouped to text-mode and graphical animations. Examples of DOS text-mode animations are the Cascade.1701.A virus, which dropped the characters on-screen to the bottom of the screen, or the Walker virus, which produced a walking man to the screen (Figure 7.4).

Graphical activation routines were rarer with early viruses, but such were found in viruses like Den_Zuk, which produced a logo on-screen

FIGURE 7.4 Activation routine of the Walker virus.

FIGURE 7.5 Activation routine of the Den_Zuk virus.

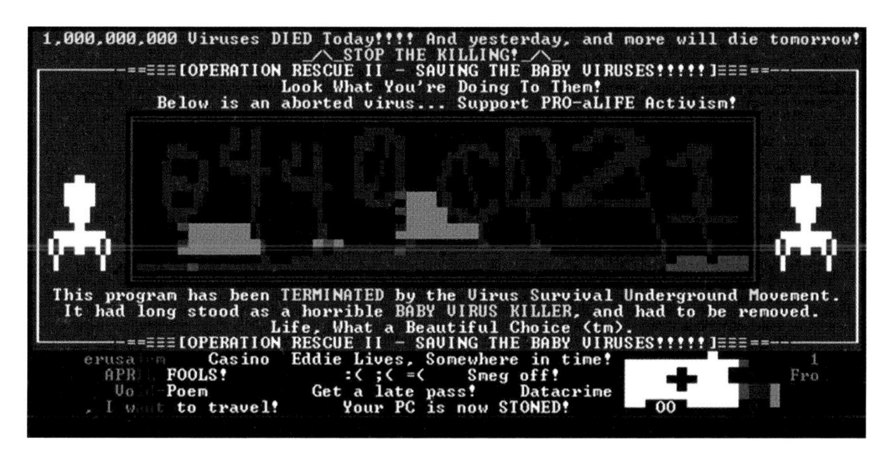

FIGURE 7.6 On-screen message shown by the Rescue virus.

(Figure 7.5), or the HH&H virus, which showed a 3D animation of a bouncing ball built out of small dots.

Messages

Viruses which displayed messages on-screen, included Stoned.Standard. A, which occasionally displayed 'Your PC is now Stoned!' Another common virus that had a message to display was the Parity_Boot.B virus, which activated by displaying 'PARITY CHECK'. A more interesting display was produced by the Rescue virus, which showed a screen full of nonsense messages (Figure 7.6).

Interactive Activations

Some early viruses stopped the PC and demanded the user of the PC to do something. For example, the Joshi virus would stop the machine on 5 January and demand the user to type 'Happy Birthday Joshi' before the machine continued working. The Casino virus forced the user to gamble in a Jackpot game with the stakes being the contents of the hard drive (Figure 7.7).

Finally, some viruses invited the user to play a game on the PC. An example of this is the Playgame virus, which displays a simple race game (Figure 7.8).

From Practical Jokes to Criminal Intent

Several viruses played practical jokes with the user. The 1989 Jerusalem. Fu_Manchu virus observed what the user was typing, and inserted comments when keywords such as 'Thatcher' or 'Reagan' were entered. The Haifa virus inserted two text lines in the middle of DOC files when they were accessed:

> OOPS! Hope I didn't ruin anything!!!
> Well, nobody reads those stupid DOCS anyway!

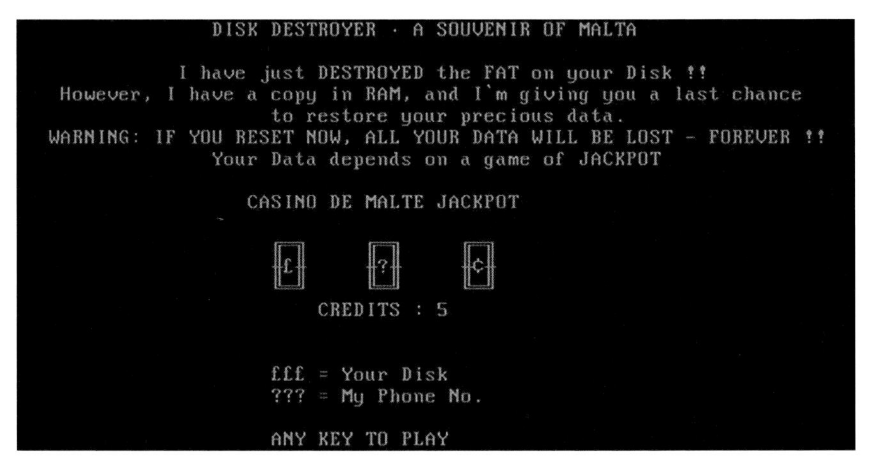

FIGURE 7.7 On-screen message shown by the Casino virus.

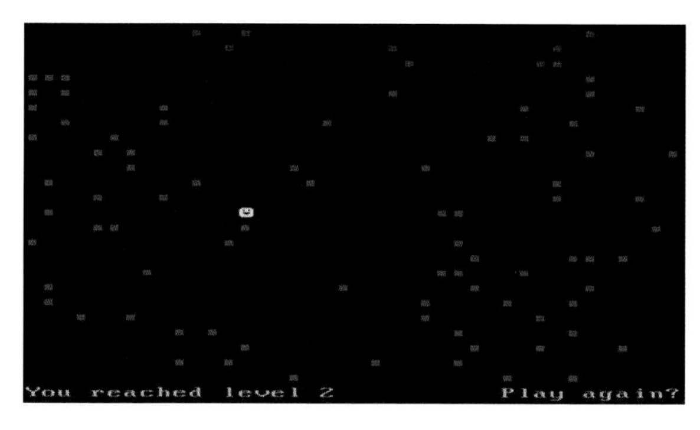

FIGURE 7.8 Playfield displayed by the Playgame virus.

Denial of Service

Some viruses just tried to make the machine unusable. Viruses that overwrote hard drives obviously did this, but then there were viruses like Monica, which set the computer's BIOS boot-up password to 'monica'.

Virus Generators – Hackers Becoming Professional

After the early viruses, things started getting more organised with virus-creating toolkits. These were written by virus hobbyist groups and could be used to produce simple viruses with no expertise needed at all. VCL (Virus Creation Laboratory) was the first virus-making program with a graphical user interface. Toolkits like the mutation engine 'MtE', made by a Bulgarian virus writer 'Dark Avenger', made self-encrypting malware accessible to any programmer.

Viruses in the Age of Windows

As PC users shifted from early DOS systems to Windows, so did malware. First Windows virus 'WinVir' was found in 1992. However, for a long while the most serious Windows attacks were not done with binary malware but with document-infecting macro viruses like Concept and Laroux. Macro viruses would use the programmable macro language (or later the VBA scripting language) to infect Word, Excel or Powerpoint documents. Such infections spread very quickly as people were sharing their documents with others. Some Excel macro viruses would slightly modify numbers inside spreadsheets on purpose. Such slow corruption would easily go undetected for a while. This did not only corrupt the data on the hard drive but often remained unnoticed over a long time leading to faulty versions being backed up as well. If this happened a restore would be very difficult or even impossible.

Some early Windows viruses would show visual effects. For example, the Marburg virus would activate by showing a series of 'stop' signs on the desktop (Figure 7.9).

These viruses were still created by hobbyists – for fame, for challenge or just because they could.

Email Worms

In 1998 we found viruses that spread via email. This quickly became the norm. The first email spreader was called Happy99. Happy99 was an email worm which claimed to be a greeting card wishing Happy New Year 1999, and would show you fireworks on your screen. While the fireworks animation was playing, the worm would copy the email address book and email itself to every address. The largest email worm outbreaks in history were Melissa and LoveLetter (aka ILOVEYOU). They both infected millions of computers, while they really didn't try to do anything other than spread.

Network Worms

After email worms, the next step in evolution was network worms. They would spread from one machine to another via direct network connection, using vulnerabilities in remotely accessible services on Windows computers (such as RPC or LSASS). In effect, they would spread very quickly over local area networks or the global internet. We saw some of the worst outbreaks in history during the year 2003: Slammer, Sasser, Blaster, Mydoom, Sobig and more. They went on to do some spectacular damage. Slammer infected a nuclear power plant in Ohio and shut down Bank of America's ATM systems. Blaster stopped trains in their tracks

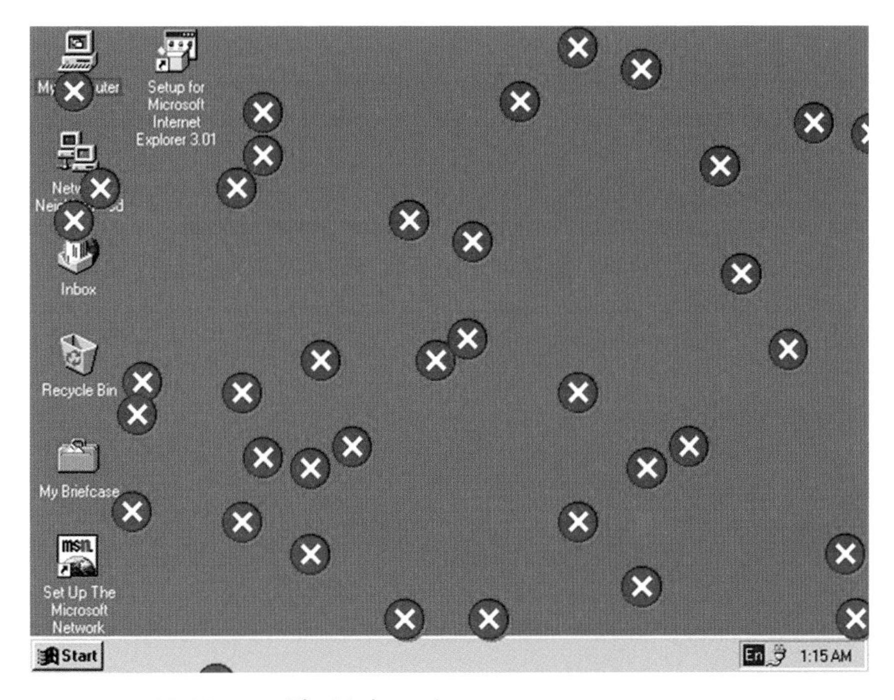

FIGURE 7.9 The Marburg virus.

outside Washington DC and shut down Air Canada check-in systems at Canadian airports. Sasser thoroughly infected several hospitals in Europe.

Commercialisation – Viruses Making Money

In 2003, we found a virus called Fizzer. This turned out to be the first PC virus written to make money. It did it by turning infected computers into spam-sending proxies. After this, many hobbyist virus writers realised the value of their skills and started making money with viruses.

We then started seeing malware with keyloggers to steal passwords and credit card numbers. Then banking trojans that would target online banking sessions. Then ransom trojans that would encrypt your files and demand a payment in order to get your files back.

Botnets

By connecting infected machines into a network that can be centrally controlled, malware writers started to operate botnets. The largest botnets in existence number millions of infected machines. Such an army of computing power can be used to do lots of things. For example, they can be used to launch very powerful denial-of-service attacks by generating massive amounts of network traffic directed at a target address online, rendering it inaccessible to customers.

Exploit Kits

As the web exploded in popularity, attacks shifted from email to the web. This is still the most common way of getting infected in 2015. Attackers break into popular websites and install an invisible 'exploit kit' on the site. As users visit such a site, their computer gets hit by various exploit codes, trying to find vulnerability in the web browser or browser extensions such as Java or Flash. In effect, Exploit Kits automate the process of infecting computers via exploits. There is no exploit without vulnerability. Ultimately, vulnerabilities are just bugs, that is, programming errors. We have bugs because programs are written by human beings, and human beings make mistakes. Software bugs have been a problem for as long as we have had programmable computers – and they are not going to disappear. But even the most serious vulnerabilities are worthless for the attacker if they get patched. Therefore, the most valuable exploits are targeting vulnerabilities that are not known to the vendor behind the exploited product. This means that the vendor cannot fix the bug and issue a security patch to close the hole. If a security patch is available and the vulnerability starts to get exploited by the attackers five days after the patch came out, the users have had five days to react. If there is no patch available, the users have no time at all to secure themselves; literally, zero days. This is where the term 'Zero Day Vulnerability' comes from: users are vulnerable, even if they have applied all possible patches.

One of the key security mechanisms continues to be patching. But for Zero Day vulnerabilities, there are no patches available.

Mining

In 2008, a mathematician called Satoshi Nakamoto submitted a technical paper for a cryptography conference[3]. The paper described a peer-to-peer network, where participating systems would do complicated mathematical calculations on something called a 'blockchain'. This system was designed to create a completely new currency: a crypto currency – in short, a currency which is based on maths. The paper was titled 'Bitcoin: A Peer-to-Peer Electronic Cash System'. Bitcoin is not linked to any existing currency – its value stems from its suitability to do instant transactions globally. Sending Bitcoins around is very much like sending email. If I have your address, I can send you money. I can send it to you instantly, anywhere, bypassing exchanges, banks and the tax man. In fact, crypto currencies make banks unnecessary for moving money around. Which is why banks hate the whole idea.

The beauty of the algorithm behind Bitcoin lies in solving two main problems of crypto currencies by joining them: how do you confirm transactions and how do you inject new units of currency into the system without causing inflation? Since there is no central bank in the system, the transactions need to be confirmed somehow – otherwise one could fabricate fake money. In Bitcoin, the confirmations are done by other members of the peer-to-peer network. At least six members of the peer-to-peer network have to confirm the transactions before they go through. But why would anybody confirm transactions for others? Because they get rewarded for it: the algorithm issues new Bitcoins as reward to users who have been participating in confirmations. This is called 'mining'.

When Bitcoin was young, mining was easy and you could easily make dozens of Bitcoins on a home computer. However, as Bitcoin value grew, mining became harder since there were more people interested in doing it. Even though the pound-to-BTC exchange rate has fluctuated widely, the fact remains that at the beginning of 2013, one Bitcoin was worth five US dollars and by early 2015 they were worth over 200 dollars. So Bitcoins quickly gained real-world value.

Today, there are massively large networks of computers mining Bitcoins and other competing crypto currencies (such as Litecoin). The basic idea behind mining is easy enough: if you have powerful computers, you can make money. Unfortunately, those computers don't have to be your own computers. Some of the largest botnets run by online criminals today are monetised by mining. So, you could have an infected home computer of a grandmother in, say, Surrey, running Windows XP at 100 per cent utilisation around the clock – as it is mining coins for a Russian cybercrime gang. Such an attack does not require a user in front of the infected computer in order to make money. Most traditional botnet monetisation mechanisms required a user's presence. For example, credit card keyloggers needed a user at the keyboard to type in his payment details – or ransom trojans needed a user to pay a ransom in order to regain access to his computer or his data. Mining botnets just need processing power and a network connection.

Some of the upcoming crypto currencies do not need high-end GPUs to do the mining: a regular CPU will do. When you combine that with the fact that home automation and embedded devices are becoming more and more common, we can make an interesting forecast: there will be botnets of embedded devices making money by mining: Botnets of infected printers or set-top boxes or microwave ovens. Or toasters.

Whether it makes better toast or not, toasters with embedded computers and internet connectivity will be reality one day. Before crypto currencies existed, it would have been hard to come up with a sensible reason for why anybody would want to write malware to infect toasters. However, mining botnets of thousands of infected toasters could actually make enough money to justify such an operation. Sooner or later, this will happen.

International Espionage, Internet Warfare and National Security

Our enemies keep changing. We used to fight the online hackers. Then the online criminals. Nowadays, we worry more and more about governmental action. We've seen governments enter the malware scene with attacks like Stuxnet, Red October or Snake. These viruses were written by government agencies to induce malfunction in certain equipment like the centrifuges used in Iran to enrich uranium. This in itself constitutes a major shift, as the hardware used in the centrifuges was considered robust against any viruses. Thus, we have to worry about future attacks being able to interfere with or disable military infrastructure but also power plants and hospitals.

A high-profile example of cyber-attacks between governments were the 2014 attacks against Sony Pictures. After Sony announced a movie called *The Interview*, North Korea protested actively because the movie plot was built around the assassination of the supreme leader Kim Jong-un. As the movie release was imminent, Sony Pictures networks were hacked and gigabytes of internal data – including several unreleased movies – were leaked to the internet. Although the actor was never conclusively identified, the US government blamed the North Korean government for the hack.

More important than simple malware attacks on strategic targets, governments started using malware for global online surveillance. But is governmental surveillance a real problem in a world where everybody seems to be sharing everything about their life with no worries? People tweet their breakfasts, foursquare their location, Facebook their dating patterns and Instagram their friends and family. For some, this is not a problem – at least it's not a problem at the moment. And this is what all these services encourage you to do, as that's how they make all their money.

But governmental surveillance is not about the governments collecting the information you're sharing publicly and willingly. It's about collecting the information you don't think you're sharing at all: Information such as the online searches you do on search engines. Or private emails or text messages. Or location of your mobile phone at any time. Surveillance like this has only been possibly for a couple of years. The internet and mobile phones made it possible. And now it's being done for that exact reason: because it can be done. But just because it can technically be done doesn't make it right.

Future Risks and Preventing Cyber Plagues

Online technologies continue to evolve at great speed, and so do online threats. It's quite clear that we will never have perfect security. However, we have seen some great advances in security technology. For example, current smartphone operating systems offer much better security to the end user compared to legacy PC operating systems. Nevertheless, the race between virus writers and virus fighters is a never-ending game of cat and mouse.

There will always be bad people and there will always be security vulnerabilities. Thus, there will always be malware. We need to be able to build our systems to be resilient enough to work, even with the continuous risk of malware – or cyber plagues.

Summary

Attacker attribution is one of the most important things an organisation can do to protect themselves. Different kinds of organisations are targeted by different kinds of attackers. And we have no hope of defending ourselves if we don't understand who the attackers are. The various attackers have different motives, they use different techniques and they pick different targets. The good news is that not every organisation is targeted by all the attackers. The bad news is that no one else can tell your potential attackers as well as you yourself. Attacker attribution work is hard to outsource. The PC malware problem has evolved from simple hobbyists to government-grade attackers. We are seeing more new malware samples today than ever before. Unfortunately, we are more dependent on computers and online services than ever before and we can already see attacks forming against the next big target: the internet of things.

Beyond that one may ask if and when computer viruses will be able to cross the boundary into the biological world. This may be either through available DNA-generating machines getting infected by computer viruses or through new interfaces with animal or human bodies. Ian Morris identifies such a transfer as a possible source between disease pools that have triggered biological pandemics in the past.

Unfortunately, it's likely that things will get worse before they get better.

References

1. Mikko Hypponen, "Retroviruses," journal article, *Virus Bulletin*, 9/93.
2. Mikko Hypponen, "Virus Activation Routines," journal article, *Proceedings of EICAR*, (1995) pp. T3 1–11. 3.
3. Nakamoto, Satoshi, *"Bitcoin: A Peer-to-Peer Electronic Cash System"*, journal article published on bitcoin.org, 1 Nov 2008.

Further Reading

Brian Krebs, *"Spam Nation: The Inside Story of Organized Cybercrime-From Global Epidemic to Your Front Door"*, published by Sourcebooks, 18 Nov 2014, ISBN 978-1501210426.

Bruce Schneier, *"Schneier on Security"*, published by Wiley, 29 Sep 2008, ISBN 0-470-395354..

Michael Sikorski, *"Practical Malware Analysis: The Hands-On Guide to Dissecting Malicious Software"*, published by No Starch Press, 3 Mar 2012, ISBN 978-1593272906.

Peter Szor, *"The Art of Computer Virus Research and Defense"*, published by Addison-Wesley Professional, 13 Feb 2005, ISBN 978-0321304544.

8 The Human Plague

STEPHEN EMMOTT

> It is now beyond reasonable doubt that *we* are the drivers of almost every global problem we face and that every one of these problems is set to grow, as we continue to grow.[1,2] Can we say that humans constitute a 'plague'. Well, it turns out that this is not too ridiculous a comparison to make. The term 'plague' is thought to originate from the Latin *plāga* ('blow'), or what we would now call an 'infection' of a host by disease-causing agents. This includes the rapid multiplication of such organisms and their actions on the host, and the reaction of the host to these organisms and the toxins they produce. I argue the analogous case; that we humans are agents who, through our rapid multiplication and our actions, are now starting to have a deleterious impact on our host, our planet. As a consequence, there is now an emerging reaction of our host to us and our actions.

Just over 200 years ago, there were fewer than a billion humans living on Earth.[3] When I was born in 1960, there were three billion.[4] In 2015, the human population has more than doubled to over seven billion people, and the UN projections expect that there will be ten billion humans on earth sometime towards the end of this century.[4] This is if, optimistically, fertility rates continue to decline as they have done over the last fifty years. It is worth mentioning that this fifty-year decline in fertility rate, which many are keen to point out as a counter-claim to the argument that we have a problem, is actually slowing. It is also worth highlighting that if the current fertility rate were to continue unchanged (which the United Nations defines as the 'constant fertility variant'), the global population at the end of this century would not be ten billion people, but will be twenty-eight billion.[4]

With this in mind, I should point out that the UN population statistics have been systematically revised upwards over the past decade.

The 'peak' population under the 'expected scenario' in 2100 was under ten billion. It is now over 11 billion according to estimates at the time of writing (Figure 8.1).[4]

In a very short period, we have fundamentally altered every single component of the life support system that our species and millions of other species depend upon for well-being and even our survival, and we continue to do so at an accelerating pace.

Land

Perhaps the most obviously visible sign of how we are altering our life support system – Earth – is our transformation of land. Nowhere is this more visible than urbanisation. As of 2014, over 50 per cent of the human population lived in cities.[5] By 2050, urban population is projected to grow to about 66 per cent.[5]

Some of the 'big cities' of the nineteenth and twentieth centuries such as London (population mid 20th Century 6 million now approximately 8 million) and Paris, are now being eclipsed in size by new, rapidly growing

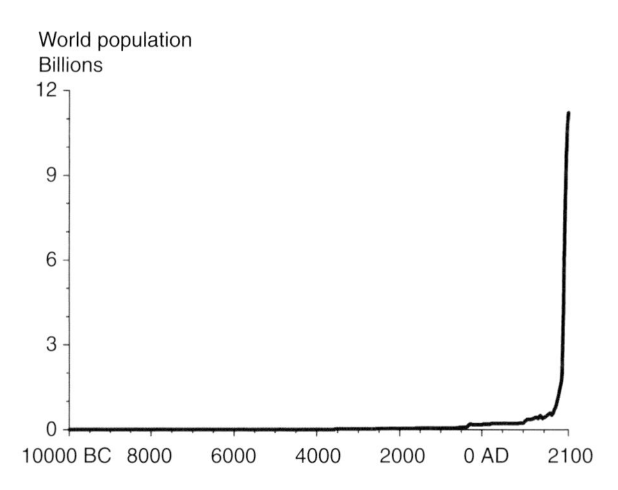

FIGURE 8.1 World population growth over centuries in billions from 10,000 BC projected to 2100 AD.
Figures from 1950 from United Nations Dept. of Economic & Social Affairs.

metropolitan areas or 'mega-cities' across the world. These include Lagos, Nigeria, population 21 million (population in 1960 0.6 million) and rising rapidly; Sao Paulo, Brazil, population 21 million; Shanghai, population 34 million and Karachi, population c.23 million (population in 1960 estimated to be 0.5 million).[5]

Food

The greatest use of land by humans is for agriculture. Approximately 40 per cent of the entire ice-free land surface of the planet is now used for agriculture.[6,7,8] (the majority of the remaining land is either desert, savannah, protected areas and cities and towns [areas not suitable for agriculture], or the world's remaining forests.) Demand for food is expected to double from 2005 by 2050 as a consequence of the combination of population increase and changing diets.[9]

The 'rational optimists' (a self-description for those people who assert that technological progress and human ingenuity will continue to enable us to solve the big problems of this century) readily make the claim that this is not an issue, and we will be easily able to achieve this without significant additional land use because the 'Green Revolution' will continue unabated throughout this century. The Green Revolution is the term used to describe the scientific and technological developments such as pesticides, herbicides and fertilisers, combined with new high yielding crops, which have greatly increased global food production between 1950 and now. We have undoubtedly benefited greatly over the past fifty years of the Green Revolution, in terms of enabling more of the growing human population to be fed. Indeed, since the 1950s, the Green Revolution has increased crop yields every decade, with little increase (between 5 and 20 per cent) in land use.[10]

But the Green Revolution has come at a very hefty price. We have had to *buy* every extra unit of food with massive quantities of chemicals, energy and water.

Furthermore, there are good reasons that we cannot blindly assume, as the so-called rational optimists do, that the Green Revolution will continue to provide us with more and more food for the next sixty years as it has for the past sixty years. Indeed, there are many reasons why we should be extremely concerned about food security. First, in many agriculturally intensive areas of the world there is a rapidly growing vulnerability to soil erosion and desertification as a result of the very intensive agricultural practices which gave us the Green Revolution, and as a result of climate change.

Second, in many of the same areas, and in areas of rapidly growing population (and thus increasing food demand), there are significant risks to future water security due to depletion of groundwater, also as a result of the same intensive agricultural practices, and increasing demand for water by humans. It is worth mentioning that of all the water on Earth only about 2 per cent of it is readily usable by humans. Of this 2 per cent approximately 70 per cent is used for agriculture.[11,12]

Third, crops are highly vulnerable to extreme weather events, from extreme wet to extreme hot, that either shut down or irreparably damage the photosynthetic and flowering (or 'grain filling') machinery of plants, significantly reducing or even wiping out yields altogether. Notably, extreme weather events are on the increase due to climate change.

Fourth, a significant proportion of the world's crops are increasingly susceptible to a range of (especially fungal) pathogens. We needed to breed-out crop diversity to create high-yielding crops, but reduced genetic diversity in most of the world's staple crop types make plants susceptible to pandemics just as much as they do in animals (such as cheetahs, described in Chapter 5 by Stephen O'Brien). Fighting this susceptibility with chemicals, which the green revolution has relied upon so heavily, is likely to become less efficient with an increasing resistance to our 'arsenal' of fungicides caused by the heavy usage of exactly those as part of the Green revolution.[10]

In short, global food security is rapidly emerging as a major issue.

Energy

A lot of energy is required to produce, transport, use and dispose of all the things we consume. Most of this energy has, and will continue, for the foreseeable future, to, come from oil, coal and gas. The US Energy Information Administration and the UK Government both estimate that energy demands are likely to increase by 300 per cent by the end of this century, and the likelihood is that the principal source of that energy will continue to be fossil fuels.

Consumerism

We cannot ascribe both the impressive and what I believe to be a deleterious transformation of our planet simply in terms of the growth in the human population. It largely is a consequence of our *behaviour*. In particular, so much of our culture now revolves around *consumption*; our insatiable desire to consume more and more 'stuff': televisions, iPads, iPhones, handbags, high-street 'couture', holidays, burgers, sushi, burritos, restaurant food, cameras, cars. This consumptive behaviour is at the heart of the damaging nature of the 'human plague'.

Climate

We are just starting to see the first signs of the profound impact that our activities, the combination of population and behaviour, are having on our life support system – otherwise known as Earth.

The first is on our climate. Our activities since the industrial revolution have resulted in a significant increase in atmospheric carbon dioxide concentrations. Atmospheric CO_2 concentrations have now reached 400 parts per million.[13] This means that we and every other living thing on Earth are now experiencing a level of atmospheric CO_2 that has not been present on Earth for several million years. How the planet's plants, especially the world's forests, which play a major role in the Earth system regulation, will respond to this is completely unknown.

One of the initial consequences of this level of CO_2 is a rise in global average temperatures. Since the start of the industrial revolution, the global average temperature has risen by around 1°C, (as illustrated in Figure 8.2 by the solid black line [five year average], grey line is annual average change). The Inter-Governmental Panel on Climate Change (IPCC) has stated that the world needs to limit global average temperature increase to no more than 2°C.

To achieve that, we would need to limit atmospheric CO_2 concentrations to around 425–450 parts per million.

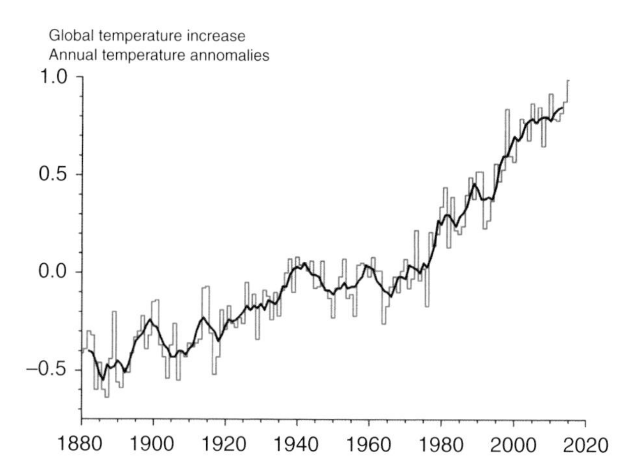

FIGURE 8.2 Global average temperature increase from 1880 to 2012. Grey line denotes annual temperature anomalies from land meteorological stations; black line shows the five-year average.[14]

Based on our current trajectory ('Business as Usual'), that goal is not going to be met. Indeed, on our current trajectory, we are heading for a catastrophic 4°C increase in global average temperature, and even a 6°C rise cannot be dismissed. Moreover, many scientists are increasingly of the view that we would need to limit CO_2 concentrations to less than the 450 ppm if we were to meet the 2 degree target. That now looks seemingly impossible.

Climate change deniers have emphasised the apparent phenomenon (yet to be properly confirmed) that there has been no change in the global average temperature for over a decade, and claim that therefore everyone who says we have a problem is just scaremongering. We should treat this claim with quite some circumspection. First of all, it is not in fact certain that *global land surface* temperature has not changed over the past fifteen years. Though, if this does turn out to be true, the scientific community does need to be able to explain it. That said, it is also worth saying that it does not need to explain it any more than having to explain a number of similar 'hiatus' in global temperature increases over the past 100 years or so. Moreover, ocean temperature has increased significantly over the same period (Figure 8.3). The reasons for this we do not yet fully understand. But what is generally accepted is that this in itself is starting to cause problems of thermal expansion, and more worryingly, ocean acidification.

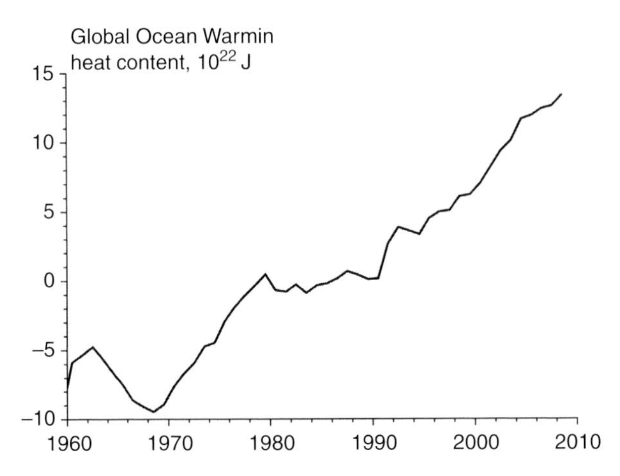

FIGURE 8.3 Global ocean warming. Data show global ocean heat content for 0–2 km depths.[15]

A briefly discussed but prominent phenomenon associated with changing global temperature is, as mentioned when discussing food in this chapter, an increasingly significant increase in extreme weather events associated with climate change. Evidence of this over the past decade has been abundant, in the UK, France, Spain, Brazil, Portugal, Greece, the Ukraine, the US, India, Pakistan and Australia, to name just a few countries, in terms of the rise in extreme heatwaves, extreme droughts and extreme floods. This has led to significant loss of crops over the last decade and is predicted by the FAO (The Food and Agriculture Organization of the UN), the US Department of Agriculture and many in the science community, to get much worse.

Ecosystems

If we never produced another molecule of carbon dioxide, we still face a major problem. Why? Because we are degrading ecosystems globally. Ecosystems provide 'ecosystem services', and chief amongst those are our food, water and our climate. We're degrading ecosystem structure and function almost everywhere on Earth. As just one example to illustrate this problem, one of the key components of the Earth's ecosystems we depend upon for our survival in significant part is the global carbon cycle.

One of the key components of the global carbon cycle is the world's biota, and, principally, the world's plants, especially forests. We are modifying dramatically every component of this critical life support system, not least through our modification and degradation of the composition of the world's forests through deforestation and other land use changes.

Loss of habitat is the principal source of terrestrial ecosystem degradation. This is a consequence of our transformation of the earth's land, through all the means mentioned above: urbanisation, agriculture, mining, transport infrastructure. Add to that over-fishing and ocean acidification as major sources of degradation of marine ecosystems. The percentage of the world's marine ecosystems fully exploited, over-exploited or collapsed (or near to collapse) has gone from around 4 per cent in 1900 (though note, only since 1950 has reliable data been collected), up to as much as 80 per cent today.[1,16]

We are degrading ecosystems, and, as a consequence, ecosystem structure and functions vital to our well-being as a species, almost everywhere on Earth. Indeed, there is a strong case to be made from recent research that we've almost certainly now embarked on the greatest mass extinction of life on Earth since the event that wiped out much of life on Earth 65 million years ago.[17]

I also wish to highlight another emerging problem for us, and one that is also a problem largely of our own making. As a consequence of increasing urbanisation, the increasing rate at which we're interacting with wildlife and our proximity with pigs and poultry, plus the rapid growth in transportation of ourselves and other animals, we are creating the potential for an unprecedented and very serious *real* plague, a global pandemic as a consequence of a novel pathogen which crosses the species barrier. This is discussed in more detail by other contributors to this book (Chapter 4, A. McLean and Chapter 5, S.J. O'Brien). Suffice to say that many epidemiologists are now of the view that a global pandemic is a case of 'when', not 'if'. It could be tomorrow, or in nine weeks' time, or in eighty-three years' time, but if it were to happen, given our population and given the rate at which transmission can occur through air travel, there is a real possibility that this could result in a major health and societal problem. The scale is unknown. But it could be over a billion people who would die from such a 'plague'. The last major global pandemic – the Spanish flu pandemic of 1918–19 – killed what was originally thought to be 50 million people (and curiously the healthy and the strong, not the weak and the infirm) and is now thought to have killed about 100 million people. This is before the innovation of budget airlines and the incredible expansion of air travel more generally, which now enables millions of us to travel across the world daily. A real problem here is that we have no way of being able to predict, prevent or manage such a plague if one were to happen, and that remains a very real and potential threat.

(In)Action

What are we doing about all the problems outlined above? Very little. As individuals, some of us probably recycle our *Observer* and our (undoubtedly carefully washed) tins of (probably organic) tomatoes. Some of us might have even bought a 'hybrid' car. But the fact is that these actions are woefully insufficient – I would even say irrelevant – given the scale and the magnitude of the problems we face.

What, then, if individual actions look likely to remain inadequate or even irrelevant, about governments and the agencies that have been set up by governments to tackle these kinds of problems? Let's just take three of some of the most important agencies in terms of their remit. The first is the United Nations framework convention on climate change (UNFCCC), whose job has been for nearly a quarter of a century to tackle the problem of climate change. The second is the UN convention to combat desertification (UNCCD), whose job has been for almost a quarter of a century to halt the rate at which desertification is occurring. The third is the Convention of Biological Diversity (CBD), whose job has been for almost a quarter of a century to halt the rate of loss of biodiversity. All of these organisations have completely failed. All we've had from governments and non-governmental agencies in thirty years is words and inaction.

What we do like to do is congratulate ourselves on our ludicrous ideas of progress. I was struck recently by a picture of a 'celebrity' congratulating a group of well-meaning people in incredibly opulent surroundings for being 'champions of the Earth'. Absurd.

What about actual political leaders? While the faces change periodically, the lack of real action remains the same. We have had thirty years of words and inaction, and, despite summits in Copenhagen, Warsaw, Brisbane and no doubt the Paris climate summit, it is now increasingly clear that we can expect another thirty years of the same from our global leaders.

All the while we're heading deeper and deeper into trouble.

Conclusion

In keeping with the *Plague* theme of this book (I've previously used the analogy of an asteroid colliding with Earth, but 'plague' also illustrates the point), it is worthwhile to consider that if we learned tomorrow that scientists had discovered that a deadly new virus existed in South East Asia, that had a 50 per cent probability of wiping out 70 per cent of all life on Earth, we would collectively organise ourselves into unprecedented action. Every government would marshal every university and every business to find a way to stop it, or find a way to save the maximum number of people on earth if option one, preventing it, couldn't be achieved.

I think we're in almost precisely that situation now, except that there isn't a real plague (at least not right now) and there isn't a date. The problem is *us*.

We urgently need to do something. And it needs to be something radical because we can rightly call the situation we face right now an unprecedented planetary emergency.

References

1. Steffen W., Grinevald J., Crutzen P.J., McNeill J. (2011) The Anthropocene: Conceptual and historical perspectives. *Philosophical Transaction of the Royal Society* 369:842–67. DOI: 10.1098/rsta.2010.0327.
2. Rockström J. et al. (2009) Planetary boundaries: Exploring the safe operating space for humanity. *Ecology and Society* 14(2). Retrieved from: www.ecologyandsociety.org/vol14/iss2/art32/. Accessed 5 May 2015.
3. U.S. Census Bureau (USCB). (n.d.). *World Population: Historical Estimates of World Population.* Retrieved from: www.census.gov/population/international/data/worldpop/table_history.php. Accessed 5 May 2015.

4. Population Division of the Department of Economic and Social Affairs of the United Nations Secretariat. *World Population Prospects: The 2012 Revision.* Retrieved from: http://esa.un.org/unpd/wpp/index.htm. Accessed 5 May 2015.
5. Population Division of the Department of Economic and Social Affairs of the United Nations Secretariat. *World Urbanization Prospects: The 2014 Revision.* Retrieved from: http://esa.un.org/unpd/wup/. Accessed 5 May 2015.
6. Haberl et al. (2007) Quantifying and mapping the human appropriation of net primary production in earth's terrestrial ecosystems. *Proceedings of the National Academies of Science* 104(31):12942–7. DOI: 19.1073/pnas.0704243104.
7. Foley et al. (2005) Global consequences of land use. *Science* 309:570–4. DOI: 10.1126/science.1111772.
8. Foley et al. (2007) Our share of the planetary pie. *Proceedings of the National Academies of Science,* 104(31):12585–6. DOI: 10.1073pnas.0705190104.
9. Tilman et al. (2011) Global food demand and the sustainable intensification of agriculture. *Proceedings of the National Academies of Science* 108(50):20260–4. DOI: 10.1073/pnas.1116437108.
10. Fisher et al. (2012) Emerging fungal threats to animal, plant and ecosystem health. *Nature* 484:186–94. DOI: 10.1038/nature10947.
11. UN World Water Development Report 4 (WWDR4). Retrieved from: www.unesco.org/new/en/natural-sciences/environment/water/wwap/wwdr/wwdr4-2012/. Accessed 5 May 2015.
12. Hoekstra, Mekonnen (2012) The water footprint of humanity. *Proceedings of the National Academy of Sciences* 109:3232–7. DOI: 10.1073/pnas. 1109936109.
13. Earth System Research Laboratory: Global Monitoring Division. *Trends in Atmospheric Carbon Dioxide.* Retrieved from: www.esrl.noaa.gov/gmd/ccgg/trends/. Accessed 5 May 2015.
14. Data from NASA GISS Surface Temperature (GISTEMP) Analysis obtained from http://cdiac.ornl.gov/trends/temp/hansen/graphics.html. Accessed 5 May 2015.
15. From Levitus S. et al. (2012) World ocean heat content and thermosteric sea level change (0–2000 m), 1955–2010, *Geophys. Researach Letters* 39:L10603.
16. Lotze, Worm (2009) Historical baselines for large marine animals. *Trends in Ecology & Evolution* 24:254–62. DOI: 10.1016/j.tree.2008.12.004.
17. Barnosky et al. (2011) Has the Earth's sixth mass extinction already arrived? *Nature* 471:51–7. DOI: 10.1038/nature09678.

Further Reading

Emmott S. (2013) *10 Billion.* Penguin UK.
Dorling D. (2013) *Population 10 Billion.* Constable.
Weisman A. (2013) *Countdown: Our Last, Best Hope for a Future on Earth?* Little Brown and Company.
Malthus T. (1798) *An Essay on the Principle of Population.* J. Johnson.

9 Plague as Metaphor

ROWAN WILLIAMS

There is a real difference between describing a public health crisis as an epidemic and calling it a plague. As far back as we can go in the European tradition, words associated with 'plague' carry an extra charge; they connote something more than accident and bring into the account we give of medical disaster the suggestion of some kind of personal agency. Homer's *Iliad* famously starts with a picture of disaster, sickness striking the Greek troops under the walls of Troy, and shows us the god Apollo unleashing his arrows against animals and human beings, in revenge for an insult offered to one of his priests. In the first book of the *Iliad*, the sickness is variously called *nousos*, *loimos* and *loigos*: roughly, a disease, an outrage or injury, and a disaster or devastation – though Homer does use *plege* elsewhere, the origin of the Latin *plaga* – a blow or stroke. Simple description gives way to a mode of speaking that is not only dramatically personal but also moralised: sickness is not only connected with someone's agency but is understood as a moral consequence of events, a punishment.

At the most obvious level, speaking like this is a way of asserting meaning in a situation where we may otherwise feel helpless, at the mercy of arbitrary forces. It may not exactly be welcome to think of ourselves as receiving punishment for our misdeeds; but it makes sense of a kind. When the workings of natural processes are made personal in this way, we can imagine 'negotiating' with them: what we do or say may make a difference. Even if we are too late to avert the plague that now afflicts us, we can perhaps shorten its duration by searching out the cause and taking appropriate remedial action, and we may be better able to avoid such disasters in future. Plague understood in this sense as the stroke of a hostile agent prompts us to examine our memories, to retell our stories, so that we can discover what we ought to negotiate about and with whom. The Athenian altar 'to an unknown god' which St Paul makes so much of in the seventeenth chapter of the Acts of the

Apostles probably reflected originally not a pious reaching out towards unutterable divine mystery but the anxious attempt of someone to render proper acknowledgement to whatever divine force had brought about a particular piece of good or bad fortune, when it had proved impossible to work out exactly what piece of good or bad behaviour, in relation to which god, had caused it. It is an insurance policy, an open cheque for the business of negotiation.

To call an event a 'blow' or 'stroke', *plaga*, is to set in train both an investigative and a narrative process. It is also of course to set in motion a process designed to end or avert more disaster, more *loimos* and *loigos*. The agency of God or the gods has been invoked, since only divine power can deploy destructive resources on this scale, 'striking from afar' in the Homeric epithet for Apollo.[1] And this has been connected to the sinful or sacrilegious agency of human beings. Now is the time, once this connection has been clarified, to undertake the next stage of action and put into place sanctions against the person or persons whose agency triggered the disaster. Talking about plague thus becomes a way of identifying what needs to be purged or expelled from the suffering community; it is a metaphor in the service of a threatened social identity. It is not difficult to provide examples of this in recent times as much as in antiquity: you have only to think of the prevalence of language about a 'gay plague' that accompanied the first reported cases of HIV/AIDS, a language that still finds echoes in an uncomfortable number of cultures around the world today. That the plague metaphor is an image to think with is clear; that it is both a powerful and ambiguous image is equally clear. And without endorsing the toxic mechanisms of scapegoating that the language encourages, it is still worth wondering why the experience of lethal epidemic or pandemic sickness seems to resist a purely 'secular' account, and what might be learned from this in thinking about human freedom and human limitation. This brief and sketchy survey of some of the ways in which the metaphor has been deployed, in ancient and modern times, is meant both to alert us to the seductive appeals of a moralised account of sickness and to suggest some of the questions about responsibility and complicity that the metaphor opens up.

Although the *Iliad* is one of the first great literary narratives to give us a story of guilty agency punished by divine aggression in the form of plague, it is Sophocles who provides what most would regard as the paradigm narrative in his Theban trilogy. The Oedipus story is, of course, a cultural resource of unusual generativity; in addition to the way in which it has shaped a whole swathe of modern consciousness by way of Freud's interpretation, there is a further dimension, more to do with issues around the intersections of individual and corporate action or inaction. *Oedipus Rex* opens – like the *Iliad* – with the evocation of devastating shared suffering and the insistent, bewildered prayer for it to be removed; Oedipus, who has already shown his quasi-divine power in his rule over Thebes, is implored to act so as to save his people's lives. He has already, he tells the people, sent to the Delphic Oracle for counsel. And what comes back from the Oracle is, untypically, a clear statement that the plague is the outward correlate of inner corruption: the murder of the old king, Laius, has never been avenged, and this is what festers and infects the city's life. So 'Banish the man, or pay back blood with blood./Murder sets the plague-storm on the city.'[2] In response, Oedipus vows to be the agent of divine healing, the agent of a divine answer to prayer: he will find and expel the criminal. 'You pray to the gods? Let me grant your prayers.'[2] The murderer is the plague personified, [2] and the king's curse is pronounced against him, even if he should turn out to be one of the royal household.

The plague here is very clearly an embodied metaphor for a diseased common conscience, a shared collusion in Laius's murder; activate that conscience by identifying the murderer and the city's infection is healed as the infecting agent is cast out. But of course the piled-up ironies of Oedipus's speeches are soon to be shown for what they are; his promise to be the agent of divine healing is indeed honoured, but not as he had expected. He is the killer, he is the poison; when he identifies himself as the source of infection, and is accordingly expelled, peace and wholeness are restored (at least for the moment). The plague is dealt with when it is seen at the heart of the self. Oedipus is – borrowing a familiar trope from Derrida – the *pharmakon*, poison and antidote together. And he is able to answer the prayers of the people in the name of the gods only when he is wholly identified not with divine sovereignty but with poisonous guilt.

His overweening assumption that he can in effect do what the gods cannot do in bringing healing is turned upside down: he will do what the gods cannot when he has no claim left to godlike freedom. Or, to put the stress somewhat differently, he is unable to heal so long as he fails to know what his complicity is in the suffering of the people – which is his complicity in the long-distant murder of the old king. Plague as metaphor for social corruption begins to be healed when those most reluctant to acknowledge their complicity are forced to see themselves clearly at last.

The sin of the monarch as the cause of plague for the people is also, of course, a biblical theme, not only in the Exodus story of the plagues of Egypt, where Pharaoh's hard-heartedness is in a very straightforward way the cause of the ten disasters that overtake his country, but, rather more subtly, in the less well-known story, in the last chapter of II Samuel, of how King David brings plague on the people of Israel. He has ordered a census of the fighting men of Israel and Judah, but this is counted an offence by God and he is given a choice of punishments: three years of famine, three months of military defeat or three days of plague (II Samuel 24.13). He opts for three days of plague, because, he says, it is better to be at God's mercy than man's; and, rather as he expected, it turns out to be possible to 'negotiate' with God's agency for a reduction in the punishment – on the grounds (interestingly, in the light of Oedipus) that the guilt is exclusively his (24.17). It is almost as though the end of the plague comes when, as in Thebes, the ruler takes full responsibility for the disaster that has arrived. And the plagues of Egypt unfold relentlessly because Pharaoh constantly denies this responsibility: he continues to oppress and murder God's children until God kills *his* child (Exodus 4.22–23), and he does not identify himself as the source of the devastation inflicted, so that it continues to affect all in Egypt.

God's agency in bringing plague is thus seen both as terrible and in one sense irresistible, but in another as 'negotiable': if we are able to sort out our responsibility, God will distinguish between innocent and guilty. Plague is associated frequently with God's action in other parts of the Exodus and desert wanderings story, and the recollection of these traditions is strong in the Psalms as well as the Pentateuch (e.g. Pss.78, 106); in the archaic hymn preserved in chapter 3 of Habakkuk, God's manifestation in the flames of lightning is accompanied also by plague and pestilence. And the same narratives are referred to or evoked in Christian Scripture to make the same dual point that God is responsible for plague and that he can be negotiated with and persuaded to be merciful (I Corinthians 10.1–13, 11.30); and the Revelation to John is perhaps the most dramatic example of the full-blown ancient tradition in Christian shape. It is perhaps worth noting in passing that the two recorded remarks of Jesus that have any bearing on the question of correlating sin and disease or disaster are entirely sceptical about such a correlation; it is notable how rapidly and comprehensively this seems to have been forgotten. But that is another story.

Thus we have, in both biblical and classical sources, a strongly defined template for reading plague as a matter of divine agency, normally punishment; and for seeing the possibilities of healing or averting disease as bound in with identifying where responsibility primarily lies, so that the innocent and the guilty can be properly distinguished. Response is not invariably expulsion (this is not implied in the case of King David), but there is a prima facie appropriateness about this which makes it a sort of default position in religious responses: the Girardian[3] mechanism is pervasive, even in the religious tradition in which arbitrary victimisation is most unambiguously challenged. But in the Middle Ages, we see not only crude victimage mechanisms – regularly directed at Jewish populations in times of intensified anxiety – but also, and increasingly in the later mediaeval period, corporate self-chastisement in response above all to the crisis generated by the Black Death. The creation of guilds of penitents, dedicated to public and shared acts of reparation (self-flagellation and so on), reflects a sense that no-one really knows where primary guilt lies in the face of universal and devastating disease: whatever degree of personal guilt you might acknowledge, there is a guilt that is common to all and must be expiated by all. To undertake this sort of penitential discipline is, in effect, to expel *yourself* from the society of the virtuous rather than to look for scapegoats elsewhere. And even when this communal self-punishment did not take such extreme forms, the regular use of processional litanies at penitential seasons and in times of crisis had always expressed the need for corporate acknowledgement of guilt (and also, in its processional forms, reflected the archaic sense that a whole territory needed purifying). The spread of print and the intensifying of state control over all local calendars led to a growing use in sixteenth- and seventeenth-century societies of national days of 'fasting and humiliation' imposed by law. Their language points in the same direction: everyone needs to be involved in the negotiations with God, because we cannot be clear just where the blame lies. Thus the 1662 *Book of Common Prayer* includes a prayer to be used 'in the time of any common plague or sickness' (along with prayers for 'Time of dearth or famine'), which refers to the rebellions of God's people in the desert and pleads 'that like as thou didst then accept of an atonement, and didst command the destroying Angel to cease from punishing, so it may please thee to withdraw from us this plague and grievous sickness'. It is not stated explicitly here what 'atonement' is now being offered, though we can assume that it is primarily the appeal to Christ's sacrifice, embodied in the penitential activities of the faithful. The same identification with the rebellious Israelites can be seen also in the service approved for use after the Great Fire and Great Plague of London: 'we are that incorrigible nation who have resisted thy judgments, and abused thy mercies.'[4] In other words, whatever people believed about the precise origins of a disastrous epidemic, the remedy was for all to accept their complicity and show this through shared penitence. The stroke or blow delivered by God in the shape of plague is thus a metaphor that can serve both exclusion – the search for the culprit to be expelled – and also, paradoxically, a kind of inclusion – an unqualified solidarity in guilt. But on both counts, it has become intertwined with the belief that external crisis requires internal examination; it is embedded in a systematically moralised picture of the natural world.

This could be chronicled in great detail; but, rather than pursue this exhaustively, I want to turn to a modern and rather different development out of this tradition, a development that uses the language and imagery of plague to make sense not so much of literal pestilence as of crisis in the moral world itself, both political and individual. Camus's *The Plague* (Figure 9.1) and Gabriel Garcia Marquez's *Love in the Time of Cholera* (Figure 9.2) both, in dramatically diverse registers, 'use plague to think with'. Like their classical and biblical precursors, they assume that disastrous infection cannot be met by passivity; but they are working with questions about the 'infection' of political will or social and

FIGURE 9.1 *The Plague*: a novel by Albert Camus

individual honesty rather than literal disease. Plague is a metaphor not for divine agency but for the unintended consequences of human agency (or indeed, *lack* of agency): it is a matter of grasping the ways in which we 'naturalize' circumstances that in fact we have in some sense created, and finding strategies for imagining those circumstances afresh. So, while

Camus in particular is overtly critical of the traditional rhetoric about suffering and punishment, he is also building on one strand in the traditional symbolic approach to plague: there is something to negotiate, and that negotiation can only happen when we have begun to ask difficult questions of ourselves.

Camus's notebooks make it very clear that his narrative is a coded account of what happened in Nazi-occupied France. 'I want to express by means of the plague', he writes, 'the suffocation from which we all suffered.'[5] In a letter to Roland Barthes, [5] he spiritedly denies that the novel represents any kind of withdrawal from history and politics (as some readers, including Barthes, had charged): in approaching the resistance to the Third Reich by way of this particular symbolism, Camus is seeking not to reduce human oppression to natural or irresistible processes, but to generalise as far as possible the experience of confronting without despair what *seems* to be irresistible. And to do this, he analyses closely the various aspects of life under tyranny. One of the most abidingly powerful aspects of the novel is its characterisations of the effects of totalitarian control. The theme of *exile* recurs, [6] the sense of being shut out from one's real home; as does the theme of how memory becomes empty or futile,[6] and the ways in which language itself is eroded or corrupted and narrowed in capacity.[6] But there is also an acknowledgement of an element of nemesis in the plague: people have taught themselves to believe that a plague-free existence is natural, and so that all things are possible. Human limitation, both in the plainest sense of mortality and in the more tangled and troubling sense of radical moral fallibility, is never absent from our world (which perhaps explains why readers like Barthes mistakenly thought Camus was some sort of fatalist). 'No-one will ever be free as long as there is plague, pestilence and famine'[6]: which means that at one level, absolute freedom is as a matter of fact impossible, and at another that it is only in the struggle against pestilence that we acquire such freedom as we have. And part of the novel's purpose is to show what such freedom might look like, in a mode drastically different from the Sartrean affirmation of pure radical liberty in human existence.

The arrival of plague reacquaints us with the truth that our fundamental condition is *danger*. And this danger can be identified at two opposite poles of moral response: we can deny the ever-present possibility of plague, or we can normalise it, pretending that the state of exile and privation of freedom and memory ('giving up what was most personal to them'[6]) is simply a fixed destiny. We can assume that plague is either impossible or normal.[6] But the hard human task is to see it as both possible and terrible, even pervasive and unavoidable, yet terrible. So, to recognise perpetual danger is to become aware that it is possible to say yes or no to the plague; we can decide to cooperate with it or to resist, to refuse to join forces with death: 'We must constantly keep a watch on ourselves to avoid being distracted and find ourselves breathing in another person's face and infecting him.'[6] Camus's ethic is one in which the inescapability of moral risk, the possibility of human corruption, is the strongest incentive to action, not an excuse for passivity. Thus he writes in his notebook for the novel, 'We should serve justice because our condition is unjust, increase happiness and joy because this world is unhappy. Similarly we should sentence no one to death, since we have been sentenced ourselves.'[5] And to fight the plague is – as Rieux, the physician in the novel, shows – to fight God: battling against the force of resignation to death is the essence of the secular holiness that belongs to Rieux as he argues with the Jesuit Fr Paneloux. The two sermons of Paneloux in the novel represent two kinds of religious response to the plague, both of which are to be resisted in Camus's eyes. Paneloux initially recommends moving 'towards the silence of God' in an acceptance born of the recognition that the plague is a divine chastisement[6]; but in his second sermon,[6] this is reduced to a sheer Pascalian clinging to faith, believing everything or denying everything, amid the wreckage of all explanations. For Camus, both explanation and the refusal of explanation have to be abandoned: the almost invisibly fine line between them is where the human conscience has to live, refusing illusion and resignation alike.

And this is, Camus surprisingly says,[5] an answer to the question of how we live 'without grace'. It is an ambiguous phrase: does it mean, 'How can we live when grace is not given in any recognizable form?' (how do we live in an irreversibly secular frame?) or 'How can we possibly live unless we have a register that allows us to speak of grace?' (even in an irreversibly secular frame). The second seems to be closer to what the novel and the notes imply: we have to 'do what Christianity has never done: concern ourselves with the damned'.[5] The damned are, presumably, human beings who are inescapably infected with the seeds of corruption and death but are constantly and compulsively forgetful of their state. Like Oedipus, they cannot see that the infection lies in their own pretence to be godlike, invulnerable to these things. But the non-Oedipodean response of Camus is that neither prayer nor purifying expulsion can be an adequate answer, only the mixture of struggle and self-knowledge. To know one's liability to plague is a condition of adequate struggle against it. Only when we know the inevitability of plague are we free to act as we ought. It is a starkly tragic framework for ethics, and it is not difficult to see why other French radicals found it unpalatable. Ethics, for Camus, is what happens in the face of an irresistible fate that must be resisted; because, paradoxical as it seems, *not* resisting death is a more deeply untruthful mode of human living. Not resisting is not noticing, and so living in that wholly illusory freedom which Camus describes at the beginning of the novel and which returns inexorably at the end. 'They calmly denied that we had ever known this senseless world in which the murder of a man was a happening as banal as the death of a fly.'[6] It is as though 'grace' could only be conceived in such a context as honesty, as the arrival of the unwelcome truth that we must both accept the world as it is and that 'we must . . . be mad, criminal or cowardly to accept the plague.'[5]

Camus's metaphorical plague is involved in complex conversation with the classical and biblical framework. The notebooks show that he had collected biblical passages presenting plague as judgement,[5] and he treats Paneloux as a serious interlocutor in the narrative, not as an easy target. For him, plague is, just as it was for Sophocles or the Psalmist, an event that imposes radical self-questioning and restores the possibility of human agency. But it does so not by imagining a negotiation with impersonal processes but by presenting a deceptively simple choice: are you on the side of the plague or not? Is your will ultimately a will for death or life? The inevitability of death is neither here nor there: to accept it at one level as universal and unavoidable is perfectly compatible with saying at another level that the essence of human identity is not to *desire* death. Religious and political totalitarianisms are, for Camus, an expression of the desire for death, the desire to stop desiring and to stop acting, resigning agency to others. And the crucial question is not, as in Homer or Sophocles, where blame lies for incurring divine wrath, but what can be done to stem the spread of infection: what can be done for the sake of honesty. This is all the grace we can envisage; but without this we cease to live humanly.

Despite its title, Marquez's *Love in the Time of Cholera* is not – on the surface – organised around the theme of plague in the way that Camus's novel is; but close reading reveals a pervasive metaphorical theme, brought into strong focus in the last episodes of the book, a focus that is easy to ignore given the sentimental readings of the novel that have been so popular (encouraged by the 2007 film version). It is not a consoling narrative; the repeated association of intense erotic passion with disease should warn us against the romantic reading. In both the main characters, the signs of passion are initially confused with the symptoms of cholera[7]; and Florentino is 'desperate to *infect*' Fermina with his own obsessiveness.[7] It is when Fermina is suspected of having cholera that her future husband, Dr Juvenal Urbino, first meets her: the doctor has lost his father, also a doctor, to cholera, and is determined to fight it wherever he can.[7] And the concluding ironic image is of the 'love boat' in which the aged Florentino and Fermina are at last united travelling under the plague flag, so as not to have to put into port.[7] Alongside these passages there are repeated references to the signs of mass death when the characters undertake long journeys: they repeatedly see corpses in the river and on the marshflats, corpses sometimes said to be of cholera victims but, we are given to understand, frequently victims of massacres in the endemic civil wars of the country ('"It must be a very special form of cholera," he said, "because every corpse has received the coup de grace through the back of the neck"'[7]). Even on the last, 'romantic' journey, 'there were no more wars or epidemics, but the swollen bodies still floated by'[7]; nothing has in fact changed, but, like Camus's citizens, people have allowed themselves to be persuaded that everything is possible and that pestilence belongs to the past.

That pestilence, the corruption of the moral world, is evoked in the way in which the frustrated Florentino, denied the possibility of marrying Fermina, replaces his obsession with her by an obsessive pursuit of sexual

FIGURE 9.2 *Love in the time of Cholera* by Gabriel García Márquez

activity, which Marquez clearly depicts as abusive and predatory, regularly involving rape. His last mistress is a schoolgirl, who commits suicide when she realises that Florentino has been reunited with Fermina, and this grim sequence of events casts a shadow over the apparent idyll of the lovers' reunion. The story ends with the lovers permanently 'quarantined' in their boat – captained by a man named 'Samaritano'; we should not miss the irony. Florentino has succeeded in 'infecting' Fermina, it seems, and they are now dangerous to others and need a Good Samaritan to lodge them and isolate them from a vulnerable population. Fermina has earlier been pronounced clear of cholera by her future husband; and it is clear that their marriage, while not a romantic ideal, is surprisingly erotically happy and satisfying. But with her husband's death she is drawn back into the potentially life-threatening ambit of Florentino's obsessions and ends up, if not exactly sharing his condition, resigned to her bondage to it.

Marquez's ironies in this novel are densely multiplied. Cholera acts as an alibi for those guilty of mass murder; the bullet hole in the back of the neck may, as we have seen, be wryly described as denoting an unusual strain of the disease. Cholera is also a stand-in for the correct diagnosis of erotic obsession; and we are steered towards the conclusion that the latter must therefore be as lethal as the disease and the bullet. But whom or what does it kill? The death of Florentino's teenage lover offers one answer: obsession makes it impossible to register the needs of another (even the actual needs of the object of the obsession, of course). But it has also been lethal to Florentino himself, whose 'diseased' and sometimes violent promiscuity has reduced his sensibility to a mixture of calculation and sentimentality. It is no accident that when he and Fermina at last find themselves together in bed, 'She searched for him where he was not, she searched again without hope, and she found him, unarmed. "It's dead," he said.'[7] The accounts of their lovemaking, when it finally happens, are fraught with ambiguities to do with death and emptiness. And Marquez leaves the reader with no very clear moral centre from which to make sense of all of this. Assuming that, very far from this being a simple tale of romantic love deferred, we have several possible moral readings. Is the disease a result of deferral or is there something intrinsically diseased about the initial obsession? Is the temptation we must confront the urge to embrace infection as a way of giving depth to a mortal condition that is otherwise unbearable? But if this is so, the price is every bit as high as the embrace of plague would be in Camus's world, a 'yes' to death, one's own and that of others, a 'yes' to the illusion of unlimited possibility. Florentino's final words, 'For ever', may suggest to the unwary a happy ending of sorts, but in fact reflect an illusion only possible in the isolation of the boat perpetually repeating its journey without arrival/consummation/resolution. We have seen throughout the narrative that death is omnipresent, disguised in various ways. The most persuasive though the bleakest reading is that the aspirations of extreme erotic passion are among the most potent disguises of death: the refusal to accept mortality properly for what it is and thus to be deceived as to what is possible for human beings. The uneven, flawed but intermittently satisfying marriage of Fermina and Juvenal Urbino is proposed as a liveable alternative to possessive violence; without denying eros, this involves negotiating with limits and resisting the mythologising of relationships. And, connecting this with the corpses in the rivers, it seems to be the restless search for absolute political control or settlement, the endlessly revived civil war, that produces massacres; as in private, so in political terms, the denial of death and the refusal of limits become lethal.

Much more could be written about the use of plague in modern fiction (not least the significance of the cholera theme in Thomas Mann's *Death in Venice*, where parallels with and differences from Marquez deserve exploring), but these discussions may serve to focus one or two themes. Since Susan Sontag's work on metaphorical characterisations of illness,[8] we have become increasingly sensitive to the ways in which we can moralise and symbolise suffering, often in a mode that makes constructive response harder. Sontag herself initially wants to strip away metaphor from our discourse, to remove the elements of mythical destiny or individual blame from how we speak of sickness; and it is not difficult to see why. Yet, as this survey has attempted to show, it is hard to disentangle sickness from metaphor: metaphorical language persistently invades the description of disease, and disease is a potent metaphor for other kinds of disorder. The transfer of the language of infection from illness to the body politic is a familiar (and usually sinister) political trope. But what I have been seeking to suggest here is that this interweaving of metaphor with how we speak of plague is not necessarily bound up only with problematic or paralysing perspectives. As we saw at the outset, the metaphor concealed in the very word 'plague' introduces the idea of agency: we are not faced with a process over which we have no control. If we can identify causal patterns, there is something we can begin to do about it. Initially, those causal patterns are to do with what has been done to provoke divine wrath, and the response is a mixture of expulsion and reformation. We have seen some reason not to take it for granted that the normative response is always scapegoating, though this is pretty pervasive: a community can represent itself as collectively alienated through rituals of collective reparation. This opens up a perspective in which plague allows us to think of a universally shared responsibility for a collective disaster; we are obliged to interrogate ourselves as well as looking for a culprit elsewhere (Oedipus's journey traces this with intense clarity).

The association of plague with divine agency thus triggers a process of recovering human agency in a situation where this initially seems impossible: if plague is the effect of divine decision, it is irresistible but not inevitable. It prompts self-questioning about how we have made it possible and how it can be averted in future or contained. The development of some kinds of religious discourse away from simple correlations of suffering and sin, the development implied at some points in the gospels, for example, 'releases' the metaphor for a wider application; it is no longer simply a matter of actual epidemic disease as an outward sign of our failure or rebellion (though as we shall see in a moment the way in which the correlation works in the case of Oedipus or David is in fact more subtle and suggestive than we might at first realise). Shared suffering or humiliation and helplessness must prompt us to ask what our complicity is in the situation – and to ask that is to step beyond being victims. This is what we have been examining in Camus and Marquez. The question of how we respond to crisis in ways that resist passivity and complicity comes to the fore, and Camus in particular sees plague – literal or metaphorical – as potentially awakening us to the ever-present risk of colluding with death. On Camus's analysis, we collude with death in two ways: by denying its omnipresence and by accepting its omnipresence as fate, as necessity. In Marquez's world – though this is far more fraught with irony than Camus's – it seems that collusion with death is a matter of clinging to a seductive but finally lethal fantasy of a 'magical' dimension in our humanity that absolves us from specific, incarnate responsibility for others; isolating ourselves as if we were cholera victims, flying the plague flag and refusing to recognise that this isolation is deadly for others as well as ourselves. Cholera in Marquez's novel, we might recall, is used by some of the characters as an ironic euphemism for the political massacres that lie in the background of the narrative, as it is also used as metaphor for the obsession of desire.

The roads lead back to Oedipus sooner or later. Our aspiration, our fantasy, is to be godlike, but the shattering advent of collective suffering forces us to ask how our own mortal fragility and betrayal is bound up with this suffering. The mythical world picture of Sophocles or the writer of the first Book of Samuel connects sin and suffering, but not in the simple mode of an individual punishment for individual guilt. The plague brought on by the king's sin reveals how the destiny of king and people are intertwined. The king has to see how the effects of his actions are not under his 'sovereignty'; he and his people are bound together and he is not at a divine distance. Thus Oedipus initially believes he can solve a problem that the gods have failed to solve, that he can answer the prayers to which they are deaf. And Camus shows us, at the start and the close of his novel, how we are led repeatedly to forget the reality of 'plague' and imagine that all things are possible for us, as if we were not always already infected with the possibility of infecting and being infected and could be isolated from a human condition of vulnerability and damage. Likewise, Marquez leaves us with a shockingly ambiguous depiction of the 'for ever' of romantic myth, the dream of isolation from mortal responsibility in its diversity and uncontrollable demands. Plague as metaphor is a way of insisting that we see ourselves as agents as well as victims – and as agents of death to each other when we refuse to see this. And in dramatically diverse ways, Sophocles and Camus gesture towards what Camus calls 'grace': the conviction comes into focus that to recognise complicity in this way, to name the moral risk of our shared world and the blood that is on our hands, is the beginning of whatever redemption can be imagined. Camus is committed to fighting the God who offers a short cut away from the call to honesty and the acknowledging of collusion; his enigmatic concern to speak a word to and for the 'damned' is in itself a curious piece of negative theology, worthy of longer exploration. Grace is encountered in the knowledge both that we are inextricably 'guilty' *together* and that there is an order of moral being (to put it rather awkwardly) in which death is not 'natural', not something we are bound to align ourselves with. We are guilty, we are under sentence, yet truthfulness about this points us to another frame of reference in which life has a last word. The metaphorisation of plague is about recovering the sense of agency, as we have seen, in the face of mythologies of fate and victimhood; and Camus's vision insists that this comes about when we neither normalise nor deny our condition. And that precarious balancing act between fatalism and hubris is precisely the point where human liberty and dignity belong, where grace is visible and 'plague' is seen clearly for what it is – the sheer fact of our capacity to injure and be injured, a capacity we must know if we are to live hopefully.

References and Further Reading

1. *Iliad* I.14 and passim; though the exact meaning of the epithet has been debated.
2. Sophocles (1982) *The Three Theban Plays*, translated by Robert Fagles. London:Allen Lane.
3. Girard René (1986) *The Scapegoat.* Baltimore:Johns Hopkins University Press, is the classic statement of his theory of how the spiral of 'mimetic' violence is controlled by the identifying and expelling of an arbitrarily selected victim.
4. Cummings Brian ed. (2011) *The Book of Common Prayer: The Texts of 1549, 1559 and 1662.* Oxford University Press.
5. Camus Albert (1970) *Selected Essays and Notebooks*, tr. Philip Thody. Harmondsworth:Penguin.
6. Camus Albert (2001) *The Plague*, tr. Robin Buss. London:Allen Lane.
7. Marquez Gabriel Garcia (1989) *Love in the Time of Cholera.* London: Penguin Books.
8. Sontag Susan (1979) *Illness as Metaphor.* New York:Farrar, Strauss and Giroux, and (1989) *AIDS and Its Metaphors.* New York:Farrar, Strauss and Giroux. She has some specific observations about how and why cholera was seen as a particularly significant variety of plague, largely because of the dramatic and humiliating nature of its symptoms.

Index